The Future of Digital Communication

This collection of essays explores the present and future of digital communication through a range of interdisciplinary approaches, all of which focus on the so-called metaverse. The metaverse is a combination of multiple elements of technology – including virtual reality, augmented reality, and video – where users "live" within a digital universe. The vision for this new universe is that its users can work, play, and stay connected with friends through everything. Such a vision is hinted at in existing phenomena such as online game universes.

About the Editor

Dr. Raquel V. Benítez Rojas, MAC, MBA, CMP is Program Coordinator and Professor at Faculty of Media and Creative Arts at Humber College Institute of Technology & Advanced Learning, Canada.

The Future of Digital Communication

The Metaverse

Edited by
Dr. Raquel V. Benítez Rojas, MAC, MBA, CMP
Humber College Institute of Technology
& Advanced Learning, Canada

CRC Press is an imprint of the
Taylor & Francis Group, an **informa** business

Designed cover image: Shutterstock Images

First edition published 2024
by CRC Press
6000 Broken Sound Parkway NW, Suite 300, Boca Raton, FL 33487–2742

and by CRC Press
4 Park Square, Milton Park, Abingdon, Oxon, OX14 4RN

CRC Press is an imprint of Taylor & Francis Group, LLC

Library of Congress Cataloging-in-Publication Data
Names: Benítez Rojas, Raquel V., editor.
Title: The future of digital communication: the metaverse / edited by
    Raquel V. Benítez Rojas.
Description: Boca Raton : CRC Press, 2024. | Includes bibliographical
    references and index.
Identifiers: LCCN 2023007403 (print) | LCCN 2023007404 (ebook) | ISBN
    9781032458847 (hardback) | ISBN 9781032458113 (paperback) | ISBN
    9781003379119 (ebook)
Subjects: LCSH: Metaverse. | Digital communications.
Classification: LCC TK5105.8864 .F88 2024 (print) | LCC TK5105.8864
    (ebook) | DDC 006.8—dc23/eng/20230503
LC record available at https://lccn.loc.gov/2023007403
LC ebook record available at https://lccn.loc.gov/2023007404

ISBN: 9781032458847 (hbk)
ISBN: 9781032458113 (pbk)
ISBN: 9781003379119 (ebk)

DOI: 10.1201/9781003379119

Typeset in Minion
by Apex CoVantage, LLC

Contents

Foreword

IT CANNOT BE DENIED THAT THE COMMUNICATION PROCESS has under-gone multiple and profound changes since those paintings in the Altamira caves when prehistoric men tried to indicate the appropriate place for hunting to his peers.

One of those evolutionary changes in the communication process centuries later has been digital communication, and within it a fact that has been breaking many structures strongly, such as the metaverse.

In this book, the reader will find eleven detailed studies on this communicative fact that covers much more than its initial association with lost worlds in a universe almost impossible to imagine. Conceived as a study of various aspects related to it, you can find articles ranging from the genesis of the object of study, the development of interpersonal relationships, educational, linguistic aspects, and even those related to the physical aspects of personal representations to through the avatars.

I would like to invite the reader to enjoy this investigative book with great didactic content, and to delve into its reading understanding that the metaverse is a creature with a soul and a heart that does not stop growing, for better or for worse.

Dr. Raquel V. Benítez Rojas, MAC, MBA, CMP – editor of
The Future of Digital Communication: The Metaverse.

Contributors

Alexander, Kris
Toronto Metropolitan University,
   Canada

Barrientos-Báez, Almudena
University Complutense de
   Madrid, Spain

Benítez Rojas, Raquel V.
Humber College Canada

Blanco-Fernández, Vítor
Universitat Pompeu Fabra, Spain

Caldevilla-Domínguez, David
University Complutense de
   Madrid, Spain

Deniz, Kemal
Munzur University, Turkey

Fenuta, Elizabeth
Humber College, Canada

Gałuszka, Damian
AGH University of Krakow,
   Poland, and EduVR Lab AGH

Gonzálvez-Vallés, Juan Enrique
University Complutense de
   Madrid, Spain

Guja, Jowita
AGH University of Krakow,
   Poland, and EduVR Lab AGH

Jurado-Martín, Montserrat
Universidad Miguel Hernández,
   Elche, Spain

Lachman, Richard
Toronto Metropolitan University,
   Canada

Martínez-Cano, Francisco-Julián
Universidad Miguel Hernández,
   Elche, Spain

Masłyk, Tomasz
AGH University of Krakow,
   Poland, and EduVR Lab AGH

Morales-Fernández, Beatriz
Universidad de Las Palmas,
   Spain

Patil, Dhruva
KLE, Dr. M.S. Sheshgiri College of
  Engineering and Technology,
  India

Ptaszek, Grzegorz
AGH University of Krakow,
  Poland, and EduVR Lab AGH

Saka, Erkan
Istanbul Bilgi University,
  Turkey

Ziemsen, Eva
Humber College, Canada

CHAPTER 1

Geneses, Evolution, and Metafuture

Dr. Raquel V. Benítez Rojas, MAC, MBA, CMP
Humber College Canada

CONTENTS

1.1 INTRODUCTION

In this initial chapter the geneses, evolution, and development of the metaverse will be establishing its roots and developing, proposing the key points for its future.

Meta comes from the Greek word meta (Μήταν), which means "among", "with", "after", or "beyond". When combined with words in English, meta-often signifies "change" or "alteration", as in the words metamorphic or metabolism, while verse means "universe". So, we are at a new universe in communication that the companies and social media groups are building, like Facebook, Instagram, Microsoft, or Apple that pour its resources into constructing virtual reality products and setting up the metaverse; therefore, it is a technological concept that goes beyond our current universe of neatly delineated physical and virtual worlds, where the new generations create their new digital communication universe.

DOI: 10.1201/9781003379119-1

1

The concept and idea of a virtual world as we currently understand it can be traced to the 1960s when scholars proposed a kinesthetic human-computer interface with interactive graphics, force feedback, body movements, and sound (Sutherland, 1965; Mazuryk & Gervautz, 1996). After decades of advances in computing power and networking and communication technologies, some technologists, innovators, entrepreneurs, investors, and journalists believe that the physical world and computer-generated virtual worlds are converging (Bakhtiari, 2020). The term metaverse has been in use since at least 1992. The Congressional Research Service in the United States of America conducted a literature search using the term metaverse and did not find any technical research articles in English containing the word before 1992, and currently it generally refers to the concept of an immersive and persistent virtual world where users can communicate and interact with other users and the surrounding environment and engage in social activities, like interactions in the physical world.

The metaverse is not just one type of experience. Instead, it could mean participating in a massive virtual reality multiplayer game accessed through a virtual reality headset or experiencing integrated digital and physical spaces such as location-specific immersive digital content from business users who are visiting via digital glasses or smartphones (Reaume, 2022).

Two significant drivers have catapulted recent interest and discussion:

Advancing technology is driving increased connectivity, enabling communication and immersion experiences that were not possible until recently. Information sharing, sensing, and simulation are all advancing rapidly and becoming more interconnected thanks to newfound ubiquity of high-speed advanced networking and the availability and affordability of technology to render convincing three-dimensional worlds.

People are spending more time engaging with digital systems and socializing in digital spaces than ever before. Some people are beginning to see their virtual lives as equivalent to their physical lives (Foutty & Bechtel, 2022).

The metaverse is a combination of multiple elements of technology, like virtual reality, augmented reality, and video where users "live" within a digital universe and where inklings of the metaverse already exist in online game universes such as *Fortnite*, *Minecraft* and *Roblox* (Snider & Molina, 2022).

But how will the new communication in this new universe be constructed? Are we facing a complete change on how we communicate on the internet? Do we need to create a new language? How is Web 3.0 technology going to affect our lives? How did this all start?

1.2 GENESES

As Bernard Marr (2022) explains, the key historical milestones that have led to where we are today started back in 1838 when scientist Sir Charles Wheatstone (1802–1875) outlined the concept of "binocular vision" where you combine two images – one per eye – to make a single 3D image. This concept led to the development of stereoscopes, a technology where you use the illusion of depth to create an image. This is the same concept used today in modern virtual reality (VR) headsets. Almost a century later, in 1935, American science fiction writer Stanley Grauman Weinbaum (1902–1935) published the short story *Pygmalion's Spectacles*, in which the main character explores a fictional world using a pair of goggles with sight, sound, taste, smell, and touch. But it was not until 1956 when Morton Leonard Heilig (1926–1997) created the Sensorama machine, the first virtual reality (VR) machine, which simulated the experience of riding a motorcycle in Brooklyn by combining 3D video with audio, scents, and a vibrating chair. Heilig also patented the first head-mounted display in 1960, which combined stereoscopic 3D images with stereo sound (Weinbaum, 2017).

In the 1970s, the Massachusetts Institute of Technology (MIT) created the Aspen Movie Map. This development enabled users to take a computer-generated tour of the town of Aspen, Colorado, U.S.A. The importance of this is that it was the first time a VR was used to transport users to another place. On January 1, 1983, a major event happened as this date is considered the official birthday of the internet because that's when Transfer Control Protocol/Internetwork Protocol (TCP/IP) was released, enabling computer networks to "communicate" to each other. Later, Sega introduced VR arcade machines like the SEGA VR-1 motion simulator in the early 1990s.

But it was author Neal Town Stephenson (born 1959), in his 1992 science fiction cyberpunk novel *Snow Crash*, in which he envisioned life-like avatars who met in realistic 3D buildings and other virtual reality environments, who is credited for coining the term metaverse, where humans, as programmable avatars, interact in a realistic virtual space with each other and software agents in a three-dimensional virtual space that uses the metaphor of the real world. Stephenson's metaverse appears to its users as an urban environment developed along a 100-meter-wide road, called the Street, which spans the entire 65,536 km circumference of a featureless, perfectly spherical black planet. This virtual real estate is owned by the Global Multimedia Protocol Group, a fictional part of the real Association for Computing Machinery, and is available to be bought and buildings developed thereupon. Users of the metaverse access it through personal terminals that project a high-quality virtual reality display onto goggles worn by the user, or from grainy black and white public terminals in booths. The users experience it from a first-person perspective. There is a subculture of people choosing to remain continuously connected to the metaverse; they are given the sobriquet gargoyles due to their grotesque appearance. Within the metaverse, individual users appear as avatars of any form, with the sole restriction of height, "to prevent people from walking around a mile high". Transport within the metaverse is limited to analogs of reality by foot or vehicle, such as the monorail that runs the entire length of the Street, stopping at 256 express ports, located evenly at 256 km intervals, and local ports, one kilometer apart (Stephenson, 1992).

1.3 THE START

On April 30, 1993 was when the document that officially put the World Wide Web into the public domain. Tim Berners-Lee, a British scientist at the European Organization for Nuclear Research (CERN), invented the World Wide Web (WWW) in 1989. The web was originally conceived and developed to meet the demand for automatic information-sharing between scientists in universities and institutes around the world. However, several people were already writing browsers that could handle graphics as well as text. Early browser contenders included names such as Erwise, Viola, and Midas, all for use on the X Window system, but the browser that got the most attention at the time was Mosaic. Released in September 1993, Mosaic was written by students Marc Lowell Andreessen (born 1971) and Eric Bina (born 1964) at the National Center for Supercomputing Applications (NCSA) at the University of Illinois and had the advantage of being easy to install on Unix, Mac, and Windows (Miller, 2014).

Professors Cynthia Dwork (born 1958) and Moni Naor (born 1961) in 1993 developed Proof-of-work (PoW), which is a form of cryptographic proof in which one party (the prover) proves to others (the verifiers) that a certain amount of a specific computational effort has been expended (Lachtar et al., 2020). Five years later, in 1998, computer scientist Wei Dai (born 1976), developed b-money, which was intended to be an anonymous, distributed electronic cash system. In this way, it endeavored to provide many of the same services and features that contemporary cryptocurrencies today do as well (Reiff, 2021). Dai outlines the basic properties of all modern-day cryptocurrency systems: "a scheme for a group of untraceable digital pseudonyms to pay each other with money and to enforce contracts amongst themselves without outside help" (Reiff, 2021). Although it was never officially launched. That same year, Sportsvision broadcast the first live National Football League (NFL) game with a yellow yard marker, and the idea of overlaying graphics over real-world views quickly spread to other sports broadcasting.

In 2002, Dr. Grieves, an expert in product lifecycle management, introduced the digital twin concept for designing, testing, manufacturing, and supporting products in the virtual world. Digital twins were anticipated by Yale computer expert David Gelernter's (born 1955) in his book *Mirror Worlds: or the Day Software Puts the Universe in a Shoebox . . . How It Will Happen and What It Will Mean* (1991). In this fascinating book – part speculation, part explanation – Gelernter takes us on a tour of the computer technology of the near future. Mirror worlds, he contends, will allow us to explore the world in unprecedented depth and detail. The concept and model of the digital twin was first publicly introduced in 2002 by Grieves, at a Society of Manufacturing Engineers conference in Troy, Michigan, where he proposed the digital twin as the conceptual model underlying product lifecycle management (PLM). The digital twin concept, which has been known by different names (e.g., virtual twin), was subsequently called the "digital twin" by John Vickers of NASA in a 2010 roadmap report. The digital twin concept consists of two distinct system or product parts (the physical system or product, the digital system or product) and the connections between the two systems. The connections between the physical system and the digital system include information flows and physical sensor flows between the two and their physical environment (Negri, 2017). Digital twin can be broken down into three broad types, which show the different times when the process can be used as it can be found on the TWI.

Digital Twin Prototype (DTP) – This is undertaken before a physical product is created.

Digital Twin Instance (DTI) – This is done once a product is manufactured in order to run tests on different usage scenarios.

Digital Twin Aggregate (DTA) – This gathers DTI information to determine the capabilities of a product, run prognostics, and test operating parameters.

1.4 THE DEVELOPMENT

A year later, in 2003, Second life (SL) is launched. Second Life is an "immersive, three-dimensional (3D) environment that supports a high level of social networking and interaction with information" (Glen, 2022). Second Life may be the first instance of a large-scale, metaverse-type world based on the original concept of the 3D environment. The platform principally features 3D-based user-generated content. Second Life also has its own virtual currency, the Linden dollar (L$), which is exchangeable with real-world currency (Au, 2013).

In 2006, *Roblox* is released. The gaming platform has been described as a "proto-Metaverse with a path to the Metaverse" foundational elements for the metaverse including immersive experiences, avatars that persist across games, and a digital economy (Glen, 2022). Created by David Baszucki (born 1963) and Erik S. Cassel (1967–2013) in 2004 and released in 2006, the platform hosts user-created games of multiple genres coded in the programming language Lua. This is a lightweight, high-level, multi-paradigm programming language designed primarily for embedded use in applications (Lerusalimschy et al., 1996).

It had to pass three years until a great milestone was developed in 2009 when Bitcoin became the first cryptocurrency, which transformed the way we think and manage money. With an abbreviation BTC and sign ฿, Bitcoin is a decentralized digital currency that can be transferred on the peer-to-peer Bitcoin network (Calvery, 2013). The cryptocurrency was invented in 2008 by an unknown person or group of people using the name Satoshi Nakamoto (Bearman, 2017). Three people who were supposedly Bitcoin founder Satoshi Nakamoto: Dorian Nakamoto, Craig Wright, and Nick Szabo. The cryptocurrency is decentralized thus as Antonopoulos (2014) explains in his work:

Bitcoin does not have a central authority.

The bitcoin network is peer-to-peer, without central servers.

The network also has no central storage; the bitcoin ledger is distributed.

The ledger is public; anybody can store it on a computer.

There is no single administrator; the ledger is maintained by a network of equally privileged miners.

Anyone can become a miner.

The additions to the ledger are maintained through competition. Until a new block is added to the ledger, it is not known which miner will create the block.

The issuance of bitcoins is decentralized. They are issued as a reward for the creation of a new block.

Anybody can create a new bitcoin address (a bitcoin counterpart of a bank account) without needing any approval.

Anybody can send a transaction to the network without needing any approval; the network merely confirms that the transaction is legitimate.

That same year, blockchain, also known as a distributed ledger technology (DLT) initial code, is launched. A blockchain is a distributed database or ledger that is shared among the nodes of a computer network. As a database, a blockchain stores information electronically in digital format. Blockchains are best known for their crucial role in cryptocurrency systems, such as Bitcoin, for maintaining a secure and decentralized record of transactions. The innovation with a blockchain is that it guarantees the fidelity and security of a record of data and generates trust without the need for a trusted third party (Hayes, 2022). Blockchain technology was first outlined in 1991 by Stuart Haber and W. Scott Stornetta (born 1959), two researchers who wanted to implement a system where document timestamps could not be tampered with. But it wasn't until almost two decades later, with the launch of Bitcoin in January 2009, that blockchain had its first real-world application.

Then Palmer Luckey (born 1992) created the prototype for the Oculus Rift VR headset in 2010. With its 90-degree field of vision and use of computer processing power, later, new versions were developed until the virtual reality firm acquired by Facebook in 2014 announced it is shutting down its in-house VR content arm, Story Studio. Facebook acquired Oculus VR in a $2 billion deal.

In 2011, Ernest Christy Cline (born 1972), an American science fiction novelist, wrote the novels *Ready Player One*, *Armada*, and *Ready Player Two*. *Ready Player One* is a dystopian, a speculated community or society that is undesirable or frightening, science fiction franchise which depicts a shared VR landscape called "the OASIS". The first novel was released in 2011, with a 2018 film adaptation directed by Steven Spielberg, and the second novel in 2020. The franchise depicts the year 2045 as being gripped by an energy crisis and global warming, causing widespread social problems and economic stagnation. The primary escape for people is a shared

VR landscape called "the OASIS" which is accessed with a VR headset and wired gloves. The OASIS functions both as a massively multiplayer online role-playing game and as a virtual society (Grady, 2018; Cline, 2011). Based on the book, a movie was produced with a particular approach: The founder of the OASIS, James Halliday, has died, and he has left his fortune – and control of the OASIS itself – to the person who can track down an Easter egg he's hidden inside the game. To find the egg, hunters (gunters, in the parlance of the book) will need an encyclopedic knowledge of Halliday's beloved 1980s pop culture. And our hero Wade, an 18-year-old video game addict from a trailer park, is sure that he's just the man to do it. He just must find the egg before a massive corporation gets its hands on it instead, regulating away the freedom of virtual reality and ending the OASIS as Wade knows it (Cline, 2011).

Non-fungible token (NFT) emerges from a "colored coin", initially issued on the Bitcoin blockchain in 2012. A non-fungible token is a unique digital identifier that cannot be copied, substituted, or subdivided, that is recorded in a blockchain, and that is used to certify authenticity and ownership (Merriam-Webster, n.d.) An NFT is a unit of data, stored on a type of digital ledger called a blockchain, which can be sold and traded (Wilson et al., 2021). The first known "NFT", Quantum, was created by Kevin McCoy and Anil Dash in May 2014. It consists of a video clip made by McCoy's wife, Jennifer. McCoy registered the video on the Namecoin blockchain and sold it to Dash for $4 during a live presentation for the Seven-on-Seven conferences at the New Museum in New York City. Digital art is a common use case for NFTs. High-profile auctions of NFTs linked to digital art have received considerable public attention. The work entitled Merge by artist Pak was the most expensive NFT, with an auction price of US$91.8 million, and Every days: The First 5000 Days, by artist Mike Winkelmann (known professionally as Beeple), the second most expensive at US$69.3 million in 2021 (Thaddeus-Johns, 2021).

The same year, 2014, Facebook/Meta acquired the Oculus VR company for $2 billion. This was a defining moment in VR's history because VR gained momentum rapidly after this. Sony announced that they were working on Project Morpheus, a VR headset for the PlayStation 4 (PS4). Google released the Cardboard – a low-cost and do-it-yourself stereoscopic viewer for smartphones. Samsung announced the Samsung Gear VR, a headset that uses a Samsung Galaxy smartphone as a viewer (Barnard, 2022).

Two big events happened in 2015, the advent of Ethereum and Decentraland was released. Ethereum blockchain has several founders.

Vitalik Buterin (born 1994) was the one who initially published a white paper explaining the concept of Ethereum in November 2013. Following Buterin's initial work, other researchers worked in various capacities to help bring the project to fruition. Vitalik Buterin, English computer scientist Gavin Wood (born 1980), Maths whiz Charles Hoskinson (born 1987), Amir Chetrit, the most mysterious and publicity shy entrepreneur and bitcoin enthusiast Anthony Di Iorio, computer programmer Jeffrey Wilcke, the Princeton-educated computer scientist Joseph Lubin, and Mihai Alisie, who has known Buterin since 2011 when they founded *Bitcoin Magazine*, one of the first publications solely dedicated to crypto, are all considered co-founders of Ethereum. Ethereum is an open-source public service that employs blockchain technology to enable smart contracts and cryptocurrency trading without the involvement of a middleman. It hosts a notable amount of functionality for developers building solutions on Ethereum as a base. The Ethereum blockchain has a native coin that is known as Ether (ETH), which is used to pay for activity on the Ethereum blockchain. The coin also trades on crypto exchanges and fluctuates in value (Cointelegraph, 2021).

Decentraland is a digital game that mimics reality in a three-dimensional format. This 3-D, user-owned, Ethereum-based virtual reality world platform, or open-world metaverse, is a combination of virtual reality, augmented reality, and the internet. It allows users to be part of a shared digital experience in which they play games, exchange collectibles, buy and sell digital real estate or wearables for avatars, socialize, and interact with each other. It is a software that seeks to give a global network of users incentives to operate a shared virtual world, according to cryptocurrency exchange Kraken (Lodge, 2022).

In 2016, *Pokémon GO* had millions of people looking around to catch its characters from Pokéstops to Pikachu. Is one of the first games to merge digital and physical worlds, as the players must look in the physical world a digital figure, Pokémon Go It's a "real world adventure". That means it uses GPS and augmented reality (AR) to allow you to hunt and train Pokémon as you're out and about in your neighbourhood. AR functionality uses your smartphone or tablet's back-facing camera to display Pokémon as though they are in front of you (Bastow, 2016). Another two milestones in 2016 are the release of decentralized autonomous organization (DAO), sometimes called a decentralized autonomous corporation (DAC), and Microsoft's HoloLens headsets hit the market. DAO, launched by German start-up slock.it, is an organization constructed by rules encoded as a computer program that is often transparent, controlled by the organization's members and not influenced by a central government, and a decentralized version of Airbnb. DAO represents an innovation in the design of organizations, in its emphasis on computerized rules and contracts, but the DAO's structures and functions also raise issues of governance (Chohan, 2017). Microsoft's HoloLens headsets gave mixed reality (AR and VR) for the first time as it allows the creation of a holographic image in front of the user, then puts it into the real world and manipulates it using augmented reality.

Epic Games released in 2017 *Fortnite*, which is an online video game. It is available in three distinct game mode versions that otherwise share the same general gameplay and game engine: *Fortnite Battle Royale*, a free-to-play battle royale game in which players fight to be the last person standing; *Fortnite: Save the World*, a cooperative hybrid tower defense-shooter and survival game in which VR players fight off zombie-like creatures and defend objects with traps and fortifications they can build; and *Fortnite Creative*, in which players are given complete freedom to create worlds and battle. This same year, Swedish furniture giant IKEA published their innovative Place app, which allows to select a piece of furniture and view what it looks like in the home or office.

In 2020, Apple added Lidar (light detection and ranging), sometimes called "laser scanning" or "3D scanning" to iPhones and iPads, creating better depth scanning for better photos and AR. The Lidar system can detect objects at distances ranging from a few meters to more than 300 meters, but it has difficulty detecting objects at close distances. It uses eye-safe laser beams to "see" the world in 3D, providing machines and computers an accurate representation of the surveyed environment (Velodyne Lidar, 2022).

In 2021 Microsoft Mesh is presented. The tech will allow teams from different geographic locations to meet and collaborate shared AR. It is a collaboration and communications platform developed by Microsoft. It is described as "the company's ambitious new attempt at unifying holographic virtual collaboration across multiple devices, be they VR headsets, AR (like HoloLens), laptops or smartphones" (Hardawar, 2021).

Facebook changed its name to Meta in 2021, indicating its focus on shaping the future of the metaverse. Two companies also launched smart glasses (Ray-Ban Stories) or highly portable virtual reality headsets that look like sunglasses (HTC's Vive Flow).

1.5 IS THERE A FUTURE?

Now after all this development towards the integration of a 3D in the daily routine, what is the future bringing? In conclusion and based on the repot developed by Ling Zhu in 2022, analyst in telecommunications policy for the Congressional Research Service of the United States of America, the metaverse is likely to feature three key characteristics that differentiate it from two-dimensional (2D) online applications and are the key to its development: (1) an immersive, three-dimensional (3D) user experience. The concept of an immersive user experience is to provide users with an enhanced, individual feeling of presence and immersion within a virtual 3D world, expanding the human-computer interface beyond a 2D (referred to as "flat" sometimes) computer or smartphone screen; (2) real-time, persistent network access; users' experience of presence in the metaverse could be further enhanced if the virtual environment is persistent – it does not "disappear" when a user has finished using it (e.g., when the user logs off). A persistent virtual space would continue to exist and evolve even when no users interact with it. Moreover, it is available to users whenever and wherever they want. Achieving persistence would require computing and data architectures capable of hosting always on, interconnected virtual spaces as well as high bandwidth, which is the data transfer capacity of a digital communications network. In general, multi media data (e.g., graphics, audio, and video) in digital form (i.e., represented by binary codes) consumes more bandwidth than text-based data does during a data communication. A network latency indicates the amount of time it takes a network to transmit a piece of data between two nodes in the network. It could be one-way (i.e., the time from the sender to the receiver) or round trip (i.e., the time from the sender to the receiver plus the time from the receiver back to the sender). High latencies may affect the quality of some interactive services perceived by users. Latency is one of the key network performance metrics, particularly for real-time internet services such as video chat, video conferencing, and online multiplayer games; and (3) interoperability across networked platforms. Interoperability is fundamental to the internet. It is a foundational principle of the internet and could be described as what makes the internet what it is.

The future looks highly hopeful for the development of the metaverse that little by little is implemented day by day in the lives of citizens, allowing virtual business appointments, trips to unexpected landscapes from the comfort of our home, romantic dates, and, in many cases, cybersex, among others. The metaverse is here to stay and be part of anyone's life, especially the new generation that already interacts with diverse and technological worlds at a young age.

REFERENCES

Antonopoulos, A. M. (2014). *Mastering Bitcoin: Unlocking digital crypto-currencies*. O'Reilly Media.

Au, W. J. (2013, June 23). Second Life turns 10: What it did wrong, and why it may have its own second life. *Old GigaOm*. https://bit.ly/3vaZLXR

Bakhtiari, K. (2020, December 30). Welcome to hyperreality: Where the physical and virtual worlds converge. *Forbes*. https://bit.ly/3jfoJTe

Barnard, D. (2022, October). History of VR – timeline of events and tech development. *Virtualspeech*. https://virtualspeech.com/blog/history-of-vr

Bastow, C. (2016, July 11). From Pokéstops to Pikachu: Everything you need to know about Pokémon Go. *The Guardian*. https://bit.ly/3FRbHmt

Bearman, S. (2017, October 27). Bitcoin's creator may be worth $6 billion – but people still don't know who it is. *CNBC*. https://bit.ly/3YNaEww

Calvery, J. S. (2013, November 11). Statement before the United States Senate Committee on Banking, Housing, and Urban Affairs Subcommittee on National Security and International Trade and Finance Subcommittee on Economic Policy. *Financial Crimes Enforcement Network*. https://bit.ly/3YJL9MN

Chohan, U. W. (2017, December 4). *The decentralized autonomous organization and governance issues*. http://dx.doi.org/10.2139/ssrn.3082055

Cline, E. (2011). *Ready player one*. Ballantine Books.

Cointelegraph. (2021, December 28). History of ETH: The rise of the Ethereum blockchain. https://bit.ly/3Woqooq

Foutty, J., & Bechtel, M. (2022, March). What's all the buzz about the metaverse? *Deloitte*. https://bit.ly/3WCMeE5

Glen, S. (2022, March 14). History of the Metaverse in one picture. *Data Science Central*. www.datasciencecentral.com/history-of-the-metaverse-in-one-picture/

Grady, C. (2018, March 26). The ready player one backlash, explained. *Vox*. https://bit.ly/3BWpibe

Hardawar, D. (2021, March 2). Microsoft Mesh aims to bring holographic virtual collaboration to all. *Engadget*. www.engadget.com/microsoft-mesh-holograms-hololens-ar-vr-164001796.html

Hayes, A. (2022, September 27). Blockchain facts. What is it, how it works, and how it can be used. *Investopedia*. www.investopedia.com/terms/b/blockchain.asp

Lachtar, N., Elkhail, A. A., Bacha, A., & Malik, H. (2020). A cross-stack approach towards defending against cryptojacking. *IEEE Computer Architecture Letters, 19*(2), 126–129. https://doi.org/10.1109/LCA.2020.3017457

Lerusalimschy, R., De Figueiredo, L. H., & Filho, W. C. (1996). Lua – an extensible extension language. *Software: Practice and Experience, 26*(6), 635–652. https://doi.org/10.1002/(SICI)1097-024X(199606)26:6 < 635::AID-SPE26 > 3.0.CO;2-P

Lodge, M. (2022, November 22). What is decentraland? *Investopedia*. www.investopedia.com/what-is-decentraland-6827259

Marr, B. (2022, March 21). A short history of the metaverse. *Forbes*. https://bit.ly/3VkGwFS

Mazuryk, T., & Gervautz, M. (1996). Virtual reality-history, applications, technology and future. *TU Wien*. https://bit.ly/3WFJvtO

Merriam-Webster. (n.d.). Definition of NFT. *Dictionary by Merriam-Webster: America's Most-Trusted Online Dictionary*. www.merriam-webster.com/dictionary/NFT

Miller, M. J. (2014, March 12). 25 years later: How a 'mesh' turned into the world wide web. *PC Mag*. https://bit.ly/3Gh2GEZ

Negri, E. (2017). A review of the roles of Digital Twin in CPS-based production systems. *Procedia Manufacturing, 11*, 939–948. https://doi.org/10.1016/j.promfg.2017.07.198

Reaume, A. (2022, June 16). What is the metaverse? Its meaning & what you should know. *Seeking Alpha*. https://seekingalpha.com/article/4472812-what-is-metaverse

Reiff, N. (2021, October 25). What is B-Money? *Investopedia*. www.investopedia.com/terms/b/bmoney.asp

Snider, M., & Molina, B. (2022, January 20). Everyone wants to own the metaverse including Facebook and Microsoft. But what exactly is it? *USA Today*. https://bit.ly/3HXMWb3

Stephenson, N. (1992). *Snow crash*. Bantam Books.

Sutherland, I. E. (1965). The ultimate display. *Proceedings of IFIP Congress, 65*(2), 506–508. http://papers.cumincad.org/data/works/att/c58e.content.pdf

Thaddeus-Johns, J. (2021, March 11). What are NFTs, anyway? One just sold for $69 million. *The New York Times*. https://bit.ly/3BWNLwU

Velodyne Lidar. (2022, June 3). What is LiDAR? Learn how LiDAR works. https://velodynelidar.com/what-is-lidar/

Weinbaum, S. G. (2017). *Pygmalion's spectacles*. CreateSpace Independent Publishing Platform.

Wilson, K. B., Karg, A., & Ghaderi, H. (2021). Prospecting non-fungible tokens in the digital economy: Stakeholders and ecosystem, risk and opportunity. *Business Horizons, 65*(5), 657–670.

Zhu, L. (2022, August 26). The metaverse: Concepts and issues for congress. *Congressional Research Service*. https://crsreports.congress.gov/product/pdf/R/R47224

CHAPTER 2

The Metaverse in Communication

Reflections from Neuroscience

Dr. Almudena Barrientos-Báez

University Complutense de Madrid, Spain

Dr. David Caldevilla-Domínguez

University Complutense de Madrid, Spain

Dr. Juan Enrique Gonzálvez-Vallés

University Complutense de Madrid, Spain

CONTENTS

2.1 INTRODUCTION

As we say, in many ways, the metaverse already exists. We already can create virtual worlds and interact with them through our screens. The virtual reality devices that exist today, and are planned for the future,

DOI: 10.1201/9781003379119-2

are limited – in a very, very limited definition – to bringing the screen close to our eyes and closing the speakers around our ears (to the dismay of generations of parents who struggled to keep their children's faces away from the TV), with the addition of new sensors that capture the movement of the hands, eventually of the feet (both already commercially available thanks to the technology developed for video game consoles in the last 15 years), and a tool already known in neuromarketing: eye tracking. All this at a considerable price. According to Vera Ocete et al. (2015):

> This technology is still in its initial development phase, and its future development will depend on the advances that can be obtained in both software applications and hardware tools, as well as new fields of application that have not yet been studied. However, its possibilities seem countless, especially thanks to the possibility of acting simultaneously on the different senses, creating a world in which participants have great freedom of movement and interaction.

The very concept of virtual reality is elusive to define: Vera Ocete et al. (2015) worked with the definition of "a dynamic three-dimensional simulation in which users feel introduced into an artificial environment that they perceive as real based on stimuli to the sensory organs". In other words: to consider that the environment that has been created is a virtual reality, several conditions must be met, namely the following:

- Simulation: Or representation of a system with a sufficient level of parallelism with the real one or, in the case of a fictitious environment, verisimilitude. The virtual environment must be able to convince the user to suspend disbelief. Even when the context of rules applicable to this environment does not correspond to its authentic counterpart.

- Interaction: Establish a system of interaction with the simulated environment that allows the user to modify it using its interface tools with the machine. Whether conventional peripherals or motion sensors.

- Perception: Or the quality and quantity of senses that the system can stimulate with its peripherals and sensors. While the stimulation of sight and hearing is covered with current technology, a complete immersion involves the stimulation of touch, smell, and maybe even taste in the future. This is a field that, at present, is essentially speculative science fiction, aimed at the possibility of creating stimuli directly in the brain.

2.2 A FUNCTIONAL COMMUNITY IN CYBERSPACE

The basis on which our current idea of what a metaverse is, is based on the electronic entertainment market. Mainly in video games with quasi-global servers that "host" virtual extensions of territory that simulate large cities, kingdoms, continents, worlds, and small galaxies through which their users can move simultaneously and enter into relationships ranging from casual to economic and binding. Under the cover of rapidly developing technology, players create structures, form societies, and establish networks of astonishing complexity. A striking example was reported in the media after the death in the US consulate in Libya of "Vile Rat" Sean Smith: a member of the US Foreign Intelligence Service, who, in his spare time, had become perhaps the most powerful single player in the game *Eve online*, a futuristic war and economy simulator with over 400,000 players in which Smith had become the linchpin of a huge and complex alliance of individuals who in turn micromanaged vast tracts of virtual terrain (Beckhusen, 2012).

Suddenly the world learned the extent to which interactions in online games had become far more serious and complex than simple child's play: when an alliance between players is so vast and complex that it depends on a trained diplomat to hold it together. But even before this revelation, the foundations had long been laid in the virtual world: the iconic Blizzard company had to watch its flagship cooperative multiplayer title *World of Warcraft* not only fill up with people who had found a way to turn it into a business: playing for hours to obtain in-game currency, quest items, and materials for crafting mini-games that they then sold to other players at prices set by the "market". The game established auction houses that allowed most of these transactions to be done with in-game money, eliminating the need for real money. To get an idea of the level of simulation and interaction, it is interesting to read the explanation of the "classic period" of the game by blogger "PaperBirdMaster" (2012):

> Both Ironforge and Orgrimmar [N. of A.: simulated in-game cities] are well connected and in an almost central location in their

respective continents [and to a suitable transport network]. But [these] were not the determining factors to choose to live in these capitals, the most important point, in this case, was the Auction Houses [. . .].

So many advantages in these capitals caused many new characters to travel to them [. . .] the player who did not want to go through all the trouble [. . .] could talk to [Players] [. . .] who did business transporting people to those coveted places.

[When] Auction House service was "released", this fact made other cities begin to have importance [. . .].

Thus we have aspects of urban location, the concentration of services, and communications creating business opportunities exploited by players (transportation through high-level areas) in a turn of events not at all foreseen by the developers and which ironically contradicts Stanley's parable: the principle according to which freedom in virtual environments is the first thing that is simulated, given that any action developed by the player has been foreseen beforehand by the developers who have had to implement the mechanics that make such interactions possible.

Even earlier, in 1997, the famous Origin Games label allowed itself to produce a game with an isometric perspective that was also intended to become a social experiment: *Ultima Online*, a medieval-fantasy game environment in which the players themselves had to contribute to a large extent to build the game's economy. In addition, the player-versus-player mode was unlimited: there were no areas safer than the walls and the guards (human or digital) that protected them, and once a player killed another player's character, he could steal everything he had on him without further consequences. This led in the early stages to the creation of player leagues in the face of the company's passivity to "fix it": either under the premise of acting as gangs of looters or under the premise of creating authoritarian laws and raising individuals to enforce them.

The experiment perhaps reached its climax when, during a simulated invasion of the world created to keep players entertained with adventures beyond the accumulation and protection of wealth, it was found that players reacted statistically differently depending on their origin: player-characters from American servers stayed to defend the cities to the last man; those from the old continent migrated from the cities to the country-side to wait for the event to pass. And finally, the Asian players evacuated the cities in good order, regrouped, reorganized, and reconquered them. The Spanish newspaper *El País* even had the funny idea of sending a "correspondent" (Siap the leather tanner) to play the game and report her adventures in its Thursday paper.

2.3 BACKGROUND IN POPULAR CULTURE

From the point of view of perception in popular culture, the most frequently mentioned antecedent is that of the novel *Snow Crash* by Neal Stephenson (1992), in which a computer-generated parallel reality constitutes half of the lives of the protagonists and their environment. More recently, it's the film *Ready Player One* (2018), adapted from the novel of the same name by Ernest Cline (2011), whose rights were bought by Warner before it was even completed, and which similarly presents a dystopian future in which the digital metaverse OASIS is the only lifeline and escape valve for millions of people in a context of severe economic depression. In both cases, the concept is clearly defined: virtual reality devices, treadmills to simulate movement in the virtual environment, sensor gloves, and a series of philosophical questions not entirely resolved at the end of the plot: with characters happy to have saved the digital world without major concern for the plight of the real one.

However, both popular culture references have been undoubtedly successful. *Snow Crash* is an acknowledged classic of the cyberpunk genre. While *Ready Player One* has managed to bring critics and audiences together in its positive review, perhaps because it has managed to appeal to that 80s and 90s demographic that dreamed of a high-tech future and can empathize with the concept of using the metaverse as an escape valve from a reality over which they feel they have lost control.

In both fictional and projected metaverses, both of them based on virtual reality and on more basic technology, and of course those presented to us by the aforementioned literary antecedents of the concept, part of what is sold to the user is control: this is, by design, in contrast to reality, which does not allow control over how the "users" will look or who they will be.

A word (by the way, it originated in *Snow Crash* and its use has been generalized since then) that is a window to the self-perception and desires of the individual. A concept that could almost be extrapolated to that of the "world of ideas" of classical Greek philosophy. Users are given the freedom to choose these things from the start, allowing them to enjoy the experiences on offer. Allowing them to control part of the non-verbal messages received by their interlocutors in the context of the metaverse: those sent by the first visual impression, at least (Bakker et al., 2015; Staudigl, 2022). Also, as with innovative technological supports in general, the new environment will effectively capture the attention of a good part of consumers, and more effectively than traditional interactions (Almansa-Martínez et al., 2019).

2.4 NEUROCOMMUNICATION IN THE NEW VIRTUAL WORLD

The exceptional pandemic situation has created several precedents applicable to the idea of a generalization of the metaverse that are hard to ignore. Tejedor et al. (2020) cite, for example, the mass use of digital media previously never employed on such a scale to enable the continuity of educational services. A consequence of the first global pandemic of the digital age. While Sánchez Muñoz (2021) points to the survival during those years of the grandfather of the metaverse and cryptocurrencies (the online game *Second Life*) which, through his words, we find functional and healthy almost twenty years after its original launch. More recently, Cerdá Suárez and Cristófol Rodríguez (2022) emphasized the perceived importance of neuromarketing techniques in learning environments, to maximize the effectiveness and usability of those created in digital format in the heat of the global emergency Barrientos-Báez and Caldevilla-Domínguez (2019).

Paíno-Ambrosio and Rodríguez-Fidalgo (2019) speculated on the possibilities of virtual reality and other related technologies, such as 360° video, for the journalism of the future: advancing a future in which photographs could become panoramas that place the reader or media viewer at the very center of the news location, limited only by the position of the photographer or cameraman at any given moment.

New technologies have always been a lure for young audiences. Their harnessing has historically been key to the success of the communicators of their time: from writing to the printing press, through the railroad and the revolutions of radio, telephone, cinema, television, and of course the internet, social networks, and mobile devices (Barrientos-Báez et al., 2022). The use of cinema, radio, and new transport was key to the political movements of the 1930s: not only did they convey an idea of modernity, but they also did it faster and better, by more effectively capturing the public's attention and, in many cases, combined with the broadcasting of true or false messages, designed to match those the public wished to receive.

The metaverse is the epitome of this ability to capture-hold attention: we have said before that the basis of the access devices consists of bringing the screen close to our eyes and closing the headphones around our ears. Perfect isolation from any stimulus other than what the telescreen (teleglasses in this case) is showing us. In this context, it is possible to deliver stimulation in an environment of perfect focus by the spectator. A context primes for both the application and research of neurocommunicative techniques such as eye tracking. And which requires several additional sensors (Sepphard, 2022) which, provided the approval by homeowners and society, will allow for the gathering of ample data on the reactions of an individual to each stimulus. In this regard, the aforementioned unprecedented digitalization of society during the pandemic might mean greater openness to the introduction of these sensors (Salazar, 2023, January 13). The sensorial isolation has as well its own field of use in the study of psychology and in communication-related psychological research (Boyd et al., 2022)

2.5 FEAR OF THE UNKNOWN: BUBBLES AND CRYPTOCURRENCIES

If the technology we have in front of us has the potential to reinvent journalistic photography, the prospect of the metaverse being, as suggested by Mark Zuckerberg's environment, a sort of second internet becomes somewhat more digestible. Specialized authors, such as Pierce (2021), subtly point to the possibility that the whole idea promoted by the CEO of the former Facebook is just a big smokescreen. This fear would tie in with cases of other companies that have emerged with expectations created based on successful precedents, and have announced to be building on the same lines or dimensions when in reality they were much less lucrative or innovative operations than what was being sold: one of the most famous examples being the English South Sea Company which was intentionally founded to evoke the East India Company when in reality it was just a financial institution dedicated to managing British government debt with inflated share prices (Löwe, 2021).

The company was not the last to advertise itself as "the next big thing." Just as them, with their mastermind John Blunt at the helm, they wanted shareholders to believe that they were investing in an "India 2.0 company", the phenomenon of imitation has arisen and arises in the markets. And that's what the metaverse has evoked in society: the fear of a bubble or a maneuver to create a deregulated space of capital is mixed with the fear of the unknown and incomprehension on the part of society at large. The metaverse is an initiative that seems to be driven by figures whose social reputation has suffered the wear and tear associated with economic success and public exposure (Lyu and Mañas-Viniegra, 2021). Generating a rejection that has been transmitted to the project under cover of its vagueness and apparent redundancies with existing technology.

Related to the case at hand, the success of bitcoin after its shaky beginnings generated a flood of imitators, each one more extravagant than the next. And some of these imitators turned out to be fraudulent. Some of them are "Ethereum", a cryptocurrency with no use for investment, Binance Coin, which offers reduced fees to obtain it or even one created as a joke in the heat of the "crypto" fever called "Dogecoin" whose image is created from the image of a dog very popular as a "meme" on the internet. Even this joke coin has a value of 0.3 dollars per unit, and some 30 million of them are circulating on the internet (Three Points, 2022).

This perception is not helped by the perceived recession in the technology sector, whose most resounding sequel was Mark Zuckerberg's announcement of a cut in his hiring expectations for the metaverse project. Especially after announcing huge and round figures of highly specialized personnel that would be necessary to carry it out. Facebook has not been the only one, however: Google, Twitter, and almost all the big tech companies have been gradually reducing their hiring for 2022, giving for certain the much-talked-about crisis of these sacred cows of the stock market (Rodríguez, 2022).

So far, this futuristic idea of metaverse is based on the extension of the existing technological base, the accelerated development of new interface technology that is not available today with the required quality, price, and quantity, and, most notably, on the existence of cryptocurrencies. Marr (2022) wrote for Forbes:

> Metaverse and cryptocurrency seem to be concepts that go hand in hand: virtual worlds and virtual money to spend in them.
>
> Both are an integral part of what is currently being touted as "web3," the third generation of the Internet, after web1, the World Wide Web, and web2, social networks. The idea is that this version of the Internet is more experiential and engaging, involving virtual and augmented reality (VR/AR) to create immersive 3D environments.
>
> The Metaverse and cryptocurrency are separate concepts and can happily exist without each other, as we have seen with Bitcoin, which has utility in both the real world and the virtual world. And

many visions of the Metaverse, including Mark Zuckerberg's, only tangentially involve crypto and Blockchain.

However, it is clear that there is potential synergy between the two ideas.

According to Santander Bank, we speak of cryptocurrency when referring to a digital asset that uses cryptographic encryption to prove its ownership, guaranteeing the integrity of transactions. It also serves to automatically regulate the creation of new currency, circumventing the possibility of copying/counterfeiting. In all cases, these are electronic currencies, which can only be held in a digital wallet (Jiménez & Gallardo, 2022). What makes them special compared to the existing technology before bitcoin is blockchain encryption, which has begun to be implemented in general digital commerce due to its extreme security against theft and electronic intrusions. This security is based on the decentralized nature of its management. To put it simply, although it may seem incredible, it is the computerized version of double-entry bookkeeping. This limits, for example, the possibility of the same payment occurring twice.

Thus, one of the most firmly established legs of the metaverse is based on the facilitation of secure economic relations between users. This has certainly contributed to the economic vision of the project and created doubts about its most visible and eye-catching component: virtual reality. Which also seems to be its least developed and most inaccessible aspect.

2.6 CONCLUSIONS

Eight pages of deliberations are not necessary to understand that the new is scary, nor to remember that the concept of perceived value is at the basis of modern economics but also of most daily stock market scams and earthquakes. But they are – and they are still short – to try to understand how we have reached the point where the integrated global digital community is an attainable reality that, not for its utopian promises and great potential, is free of dangers, drawbacks, and obstacles. As the genre of fiction from which it is conceptually derived –Cyberpunk – it reflects the fact that the undeniable advantages of technological progress can result in a negative balance for society if they are not accompanied by appropriate legal, educational, and ethical development.

The metaverse is presented as the latest and most effective tool to develop neuroscience and its communication branch to the next level: First, because of its potential to not only capture but also retain the user's attention. And second, because of its precedents in the creation of functional virtual communities and societies in contexts where these were not even as closely linked to real society as intended in the case of the metaverse (for the players, they operated merely as ways to spend their free time, not as complements to their life or work needs). Precedents that have seen the simulation of societies taken to the closest point to reality imaginable. Allowing experiments that would not have been possible on such a scale in any other way. These have provided first-hand information on community decision-making processes, which will undoubtedly be useful when developing predictive patterns of social behavior on a medium scale.

The video game is the audiovisual art of the future, introducing interactivity with the viewer as the indisputable basis of its appeal, and advances and changes in the user interface constituting the periodic technological advances that maintain audience interest. Virtual environments sell adventure, excitement, and interactivity: They sell a life other than their own to millions of customers who wish to receive the message that such a thing is possible and make them receptive to the messages associated with that message they wish to hear/receive. The principle of gamification seeks to associate the undesirable behavior/activity for the individual to a context that compels the user to develop it voluntarily, both in the long and short term.

The metaverse intends to go one step further: in a continuation of the very idea of Facebook, it intends to bring social relations even further into the digital environment, seizing on economic relations to justify and finance the project. This will require the development of new technologies, more practical and sensitive sensors, affordable devices for households, as once were the computer, car, or radio receiver, the creation of financial products compatible with the metaverse, cryptocurrency and blockchain systems, as well as products that introduce their use and usage skills among the general public.

In short, digitized relationships enable a wide range of spheres: personal, economic, work, leisure, artistic, etc. And they are in themselves the terrain on which the present battle for the attention of the public is being fought. The success of the metaverse as a second internet will depend exclusively on its ability to create attractive interfaces, endowed with content capable of attracting users and companies to develop their business through it. It will also depend, to a large extent, on the ability of the industry that provides it to monetize it in the most creative way possible, in a context in which ideas about this particular unpleasantness are lagging woefully behind social and technological development: anchored in fees, subscriptions, advertising, and draconian restrictions on the redistribution of content or acquired rights.

This article is part of the framework of a Concilium project (931.791) of the Complutense University of Madrid, "Validation of communication models, business, social networks and gender".

REFERENCES

Almansa-Martínez, A., Van-Zummeren Moreno, G., & Haro, R. (2019). Funcionalidades de Moodle y Edmodo en las enseñanzas medias y superiores. *Revista de Comunicación de la SEECI, 50*, 87–105. http://doi.org/10.15198/seeci.2019.50.87-105

Bakker, M., Kaduk, K., Elsner, C., Juvrud, J., & Gredebäck, G. (2015). The neural basis of non-verbal communication-enhanced processing of perceived giveme gestures in 9-month-old girls. *Frontiers in Psychology, 6*(59), 1–15. http://doi.org/10.3389/fpsyg.2015.00059

Barrientos-Báez, A., & Caldevilla-Domínguez, D. (2019). Relaciones públicas y Neurocomunicación como herramientas de mejora de la imagen de marca de los personajes públicos. *Revista de Ciencias de la Comunicación e Información, 241*, 1–13. https://doi.org/10.35742/rcci.2019.24(1).1-13

Barrientos-Báez, A., Caldevilla-Domínguez, D., & Yezers´ka, L. (2022). Fake news y posverdad: relación con las redes sociales y fiabilidad de contenidos. *Fonseca, Journal of Communication, 24*, 149–162. https://doi.org/10.14201/fjc.28294

Beckhusen, R. (2012, September 12). Diplomat killed in Libya told fellow gamers: Hope I 'don't die tonight'. *Wired.* www.wired.com/2012/09/vilerat/

Boyd, L., Garner, E., Kim, I., & Valencia, G. (2022). Cognality VR: Exploring a mobile VR app with multiple stakeholders to reduce meltdowns in autistic children. In *Extended Abstracts of the 2022 CHI Conference on Human Factors in Computing Systems* (CHI EA'22). Association for Computing Machinery, Article 239, 1–7. https://doi.org/10.1145/3491101.3519742

Cerdá Suárez, L. M., & Cristófol Rodríguez, C. (2022). Un estudio exploratorio sobre el impacto del neuromarketing en entornos virtuales de aprendizaje. *Vivat Academia, Revista de Comunicación, 155*, 1–16. https://doi.org/10.15178/va.2022.155.e1391

Deliyore Vega, M. D. (2021). Redes como espacio de comunicación para la educación virtual de estudiantes con discapacidad en Costa Rica en tiempos de pandemia. *Historia y Comunicación Social, 26*(Especial), 75–85. https://doi.org/10.5209/hics.74243

Jiménez, J., & Gallardo, M. (2022, May 24). Guía para saber qué son las criptomonedas. *Blog del Santander.* https://acortar.link/XBmaCO

Löwe, K. (2021). *Die Südseeblase in der englischen Kunst des 18. und 19. Jahrhunderts.* Reimer.

Lyu, D., & Mañas-Viniegra, L. (2021). Problemas éticos en la investigación con neuromarketing: una revisión de la literatura. *Vivat Academia, Revista de Comunicación, 154*, 263–283. https://doi.org/10.15178/va.2021.154.e1351

Marr, B. (2022, April 30). Cómo el Metaverso cambia las criptomonedas. *Forbes.* https://acortar.link/VKDfOP

Moreno López, B. (2018). El uso de la pseudociencia y la experimentación en las nuevas tendencias de comunicación publicitaria como recurso persuasivo. *Revista Latina de Comunicación Social, 73*, 1428–1444. https://doi.org/10.4185/RLCS-2018-1315

Paíno-Ambrosio, A., & Rodríguez-Fidalgo, M. (2019). Propuesta de "géneros periodísticos inmersivos" basados en la realidad virtual y el vídeo en 360°. *Revista Latina de Comunicación Social, 74*, 1132–1153. http://doi.org/10.4185/RLCS-2019-1375

PaperBirdMaster. (2012, July 24). Las ciudades de World of Warcraft. *Spamchainheal.* https://acortar.link/kQps7A

Pierce, D. (2021, October 21). Mark Zuckerberg just announced the end of Facebook. *Protocol.* www.protocol.com/facebook-meta-metaverse

Rodríguez, P. (2022, July 12). Comienza la purga en Meta: la empresa insta a los jefes a denunciar a los empleados por bajo rendimiento. *Xataka.* https://acortar.link/gmiuEI

Rúas-Araújo, J., & Barrientos-Báez, A. (2020). Neurocomunicación y persuasión: algunas experiencias con VFC y LIWC. In U. Cuesta (Coord.), *Viaje al fondo del Neuromarketing.* Fragua.

Salazar, R. (2023). *Understanding the reality and power of web3.* https://shorturl.at/duLNR

Sánchez Muñoz, G. (2021). Second life: un entorno virtual para reducir la ansiedad de los estudiantes de lenguas extranjeras. *Vivat Academia, Revista de Comunicación, 154*, 1–24. https://doi.org/10.15178/va.2021.154.e1369

Sepphard, J. (2022, May 19). *What sensors are used in AR/VR systems?* www.sensortips.com/featured/what-sensors-are-used-in-ar-vr-systems-faq/

Staudigl, T., Minxha, M., Mamelak, A. N., Gothard, K. M., & Rutishauser, U. (2022). Saccade-related neural communication in the human medial temporal lobe is modulated by the social relevance of stimuli. *Science Advances, 8*(11), 1–12. www.doi.org/10.1126/sciadv.abl6037

Tejedor, S., Cervi, L., Tusa, F., & Parola, A. (2020). Educación en tiempos de pandemia: reflexiones de alumnos y profesores sobre la enseñanza virtual universitaria en España, Italia y Ecuador. *Revista Latina de Comunicación Social, 78*, 1–21. www.doi.org/10.4185/RLCS-2020-1466

Three Points. (2022, March 3). ¿Cuántos tipos de criptomonedas existen?. www.threepoints.com/blog/cuantos-tipos-de-criptomonedas-existen

Velarde, O., Bernete, F., & Casas-Mas, B. (2019). Las interacciones virtuales con personas conocidas que no son amigas. *Revista Latina de Comunicación Social, 74*, 668–691. https://doi.org/10.4185/RLCS-2019-1351

Vera Ocete, G., Ortega Carrillo, J. A., & Burgos González, M. Á. (2015). La realidad virtual y sus posibilidades didácticas. *Etic@net, 2*(2), 1–17. http://ugr.es/~sevimeco/revistaeticanet/Numero2/Articulos/Realidadvirtual.pdf

CHAPTER 3

New Linguistic Spaces in Cyberculture

The Influence of the Metaverse on the Minifiction of Social Networks

Dr. Beatriz Morales-Fernández

Universidad de Las Palmas de Gran Canaria, Spain

CONTENTS

3.1 INTRODUCTION: THE INFLUENCE OF THE METAVERSE ON HUMAN COMMUNICATION

The 21st century is characterized by a change of era previously analyzed and proposed by academics and historians, the so-called Digital Age, with the arrival of the metaverse and an *untact* (without contact) advance in artistic, cultural, and economic communication, broadly speaking, of all societies. Therefore, first of all, it is necessary to clarify these new terms

DOI: 10.1201/9781003379119-3

that influence our behavior and the way we communicate: the metaverse is an ambiguous concept when it comes to offering a totally precise definition, but the significant essence that all theories share is, based on the work *The Metaverse: The Digital Earth-The World of Rising Trends*, by university doctor Sangkyun Kim: "the representation of worlds created by humans using digital technology to transcend the real world" (Kim, 2022, p. 12).

In this sense, our forms of artistic representation, the ways of socializing and payment methods transcend, they pass from one state to another, to multiply their possibilities and take shape in new realities, which were already with us before, since digital media such as computers, tablets, mobile phones and the Internet are tools, the first three almost bodily, referential, creational, organizational, playful and monetary.

To expedite this transition, it is necessary to coexist, to understand the possibility of two lands, the analog and the digital, where in the latter a type of society that already lived with us before the appearance of the word *metaverse*, and that it intensified in the hard years of COVID-19, that is, during the calendar years 2020–21: the *untact*, defined as the interaction without being physically together at a precise moment in time. In this way, a world is made possible in which people can socialize in other ways and with more people (Kim, 2022), thus interfering in the ways in which we communicate, which implies a necessary adaptation of language, diversifying and naming new realities, as is this case. Therefore, the great introductory questions that come up for us here are the following: how does the transition to the metaverse influence human communication? Are new forms of communication between human beings generated?

Communicating means manifesting the human need. In this broad sense of the idea, the human being is an insatiable animal, since, as expressed by Sangkyum Kim: "in analog Earth, it does not matter how many structures we build, how many new products we develop, how much we travel or how many new people we meet, we can never be completely satisfied" (Kim, 2022, p. 20). Through language we have communicated those latent absences that we found despite finding what we were looking for, since uncertainty lies in the finding, generating new needs to be, feel, and do. However, in the metaverse they want to cover that search with all the possible possibilities in all the areas that human communities gestate, thus covering all the ways in which we can or could communicate, hence the fundamental part of this new concept is the noun *universe*, without underestimating the *meta-* prefix, which places the name in a new context.

Speaking of terminology, it is relevant to emphasize the names with which the human being has been characterized as a species from *Homo sapiens*, since they are summative terms to it and clarify the path to the satisfaction of the basic and elemental desires that we have as living beings: the first mentioned refers to the following definition, "the man who thinks", it is true that our first needs were covered with imagination in the face of things not previously thought of, so that the communication itself was a mere imaginative transition to gestate realities not possible until premeditated thought; but, around the 19th century, the so-called *Homo faber* appeared, who "focused on turning the results of their imagination into visible tools . . . to create various assets faster, cheaper and in greater quantity" (Kim, 2022, p. 24); it was no longer enough for us to imagine, we had to quickly create to satisfy new desires, which could only be partially calmed with human activities and interactions through pleasure and play (Kim, 2022). Well, satisfaction is focused on the achievement of challenges and objectives, that is, a path towards a goal, the development of unsatisfied desire until it calms down, that historical search that calms down in the development of civilizations to find new questions linked to new ways of being, doing, and feeling. These desires, channeled through play, configure another summative term to the identity of the human being: *Homo ludens*, a term coined by Johan Huizinga. This concept evidence how

> the act of creating rules to find pleasure in games formed the basis of the laws and regulations required for communal societies. . . . In addition, different communities showed differences in who did what, when and where during the game, and these discrepancies became the context for the formation of different cultures.
>
> (Kim, 2022, p. 26)

We currently see this in McLuhan's global village,[1] that is, in the metaverse that has been beginning for some time through our digital media: artistic representation itself, which configures cyberculture, is an interpretive game, affecting minification. The metaverse is divided into four areas: augmented reality, lifelogging, mirror worlds, and virtual worlds. In each of them we find different communicative ways of playing, that is, of finding linguistic pleasure in interpretive searches to feel, make ourselves and be different. Lastly, in 2015, through the writer and researcher Yuval Noah Harari, a last summative term emerged that is established as the culmination of so much busy search, and that adapts very well to what the metaverse intends and its possibilities of creating and organize communities: the *Homo deus*.[2] Human beings, after covering their basic needs through their imaginative, creative and playful capacities, over 200,000 years, need to communicate higher values: happiness and eternal life (Kim, 2022). This is where the transition from the physical to the digital Earth becomes necessary and urgent: because the first does not guarantee that possibility, it only dissatisfies you; in the minification itself through literary topics, such as *carpe diem*, representative of the necessary search for human satisfaction; the *Ubi sunt?* or *post-mortem love*, among others, this impossible reality has become evident: man communicates, doubtful, what awaits him after definitively closing his eyes, thus making possible literary and mystical worlds and universes where we partially satisfy that unfulfilled desire. The metaverse and its communicative way, through this recent concept of Homo deus, already breaks with the last insatiable desire. Everything changes artistically, socially, economically and culturally.

These transitions to created worlds and new essential words, which reflect real community change, are synthesized in four dimensions that will mark the central course of all the ways in which human beings will communicate linguistically in the metaverse, influencing as well in the creation and literary reading, subject that outlines this article.

The first thing that must be clear is that if reality, even converging with analog, multiplies and covers all possibilities, human ways of communicating must cover them all, even those that are not yet thought or established, then the transmission of information is in full convergence and birth, but it has two essential elements: the linguistic acceleration of texts, oral and written, since the immediacy of the system stimulates an immediate language and, therefore, simple for its mechanical comprehension. An example could be the speed of the audios in the conversations. See on social networks such as WhatsApp or Telegram, located in the lifelogging block of the metaverse; and the preference for means of communication without implication of immediate response in real time, opting for

> more efficient communication methods such as text messages, direct messages (social networks), emoticons (social networks), surveys (applications of messaging), the selection between several options (choosing food to order in a delivery service), status updates (messaging applications) and chats (online games).
>
> (Kim, 2022. p. 30)

This reflects the transition from real time to digital, that is, the optimization and adaptation of live language to a premeditated one with a reflective possibility but lacking in spontaneity. At the communicative level, this generates, according to the research carried out by Sangkyun Kim, four new dimensions of communication, affecting all the scientific and cultural disciplines of a society, which is why it influences the literary fiction that is gestating and will gestate in the metaverse. The first dimension has to do with the new reality of communicative production and listening between several senders and receivers, that is, who speaks and who listens, since now we will have four ways of exchanging information: a single sender and several receivers, see a speech by video call; a natural two-way communication between two people, where sender and receiver exchange roles, like in an online chat; a group divided into subgroups that communicate with each other and see each other in company meetings and organizational issues through applications such as Asana; and all playing the role of issuers to express opinions, see yourself on a bulletin board where a first issuer has expressed an idea or sensitivity.

These new ways of conceiving, carrying out and interpreting information affect the literariness of the messages and the literary genres that take place in the metaverse, something that is manifested later in this study.

3.2 THE CONCEPTUAL GEAR OF CONTEMPORARY DIGITAL LITERATURE: MILLENNIAL LITERATURE

New times arise along with new identities that seek adequate methods of expression for their speeches. These own spaces have metamorphosed with those that arise in the digital world, specifically in the so-called digital literature that, following the characteristics of the liquid society that we have inherited, as expressed by Zygmunt Bauman, become more flexible and adapt to the characteristics of the instantaneity, immediacy and linguistic acceleration. In this sense, there are a series of keywords that make up a large network, such as the Internet, in the conception and cultural dissemination of contemporary digital literature, also known and pigeonholed as millennial literature. The first of these concepts is that of cyberculture.

Cyberculture, also known as Internet culture, is the culture that arises, or is emerging, from the use of computer networks for communication, entertainment and electronic marketing. It is a culture born from the use of new information technologies. If we say that literature is a reflection of the historical event, in constant mutation and modification, in the words of the researcher María Luisa Lanzuela Corella: "The literary work is not an isolated event, it is a reflection, conscious or unconscious, of the social, economic and political situation of a certain historical moment" (Lanzuela Corella, 1998, p. 259). We will arrive at the central idea that cyberculture has also been introduced in literary texts as a mirror of social, spatial and cultural interaction changes. In this conceptual gear, therefore, the introduction of other literary concepts that are related to each other in this famous global village, as Marshall McLuhan spoke: minification, microtextuality and the micro-story cannot be missing.

Focusing on the first, minification, the central axis of the other two concepts, and in the words of the researcher Olivia Sébart (2016):

> Lauro Zavala, a Mexican theorist who was very interested in this new genre, emphasizes that in the last two decades interest in short stories has grown and divides them into three categories: short stories, very short stories, and ultra-short stories. . . . These ultra-short stories are also called minifictions, mini-stories, micro-stories, sudden fiction, vignettes, microscopic story, texticle, etc. The profusion of names shows the lack of definition of the genre, a result of the disinterest of theorists as well as the difficulty of classifying this hybrid genre.

The most relevant figure of this new genre that adapts adequately to liquid times has emerged from this event: Lauro Zavala, known for his work in literary theory, film theory and semiotics, especially in relation to studies on irony, metafiction and short story. For him, minifiction is the genre of the third millennium, since it adapts to the liquidity and hybridity of society, divided and, in turn, shared with a reality expanded to the virtual world. In the words of the researcher Zavala: "Minifiction may become the most characteristic writing of the third millennium, since it is very close to the paratactic fragmentation of hypertextual writing, typical of electronic media" (Lauro Zavala, 2018). But what does he mean by hypertextual writing and paratactic fragmentary? In the words of Dr. Susana Sendra Ramos (2018):

> There are numerous substantial transformations that the digital medium has produced in relation to the literary work: the very nature of the genre, the phenomenology of reading, hypertextual writing, the new media, the relationship between text and image, the protagonism of social networks, etc.

The characteristics of Zavala's minification are met, broadly speaking, in many texts of emerging contemporary literature in the Hispanic world. We are referring to brevity, diversity, complicity, fractality, transience and virtuality.

However, we must not forget that brevity in literature has already been handled under different names: aphorisms, haikus, the famous gregueries of Ramón Gómez de la Serna . . . Even as antecedents of micro-stories, we find Rubén Darío, Vicente Huidobro and José Antonio Ramos Sucre, that is, as the direct precursors. There is, therefore, a Hispano-American root in the theory of this genre of minifiction, in addition to having Julio Torri and Augusto Monterroso as referents of the genre and their famous micro-story: "The dinosaur" (1959): When he woke up, the dinosaur still It was there. Story that, by the way, fits in a single tweet.

In this way, microtextuality is consolidated as a new literary reality with its own theoretical discipline through minification, linked to its full literary product, that is, that arises from it: the micro-story.

Each concept presented can thus be linked to the other, like a hypertext link that leads us to the other, sharing a semantic essence that leads us to the same artistic and literary reality: the acceleration of discourses.

Here, we must emphasize the linguistic theory of discourse: specifically, the time of the story versus the literal time of the discourse, that is, its duration, its speed. The speed of the speech will be defined as the relationship between the duration of the story (in minutes, days, months, years) and the length dedicated to it in the text (in lines and pages). Thus, taking an equal speed as a reference, there are two forms of change (anisochrony): acceleration and deceleration.

Deceleration is understood as the dedication of a long segment of the text to a short period of history; against acceleration, which has in its essence the opposite process, that is, the dedication of a short textual segment for a long period of history.

Socially and literarily, we find the anisochrony of acceleration within cyberculture, thanks to social networks. In addition, virtuality, characteristic of Lauro Zavala's minification, encompasses hypertextuality and cybertext. This implies that getting the message requires work on the part of the user; the reader is not limited to interpreting the text but performs actions such as active choice and decision-making through the navigation options or with the constant interaction of the author with the author through forums or Instagram games/quizzes of Twitter to generate a feedback loop between the reader and the text and between the first and the author, since complicity, another characteristic of minifiction, enriches the literary discourse, the short story.

3.3 LINGUISTIC EFFECTS OF THE METAVERSE IN LITERARY CREATION. A CASE STUDY: TWITTER

Focusing on the linguistic effects of this acceleration of the texts of the metaverse in the minification of networks, and taking as an example the previously mentioned social network Twitter, we will make a qualitative analysis following the literary and theoretical reference of Andrés Neuman and his *10 micronotes on the short story* (2012) in order to confirm the hypothesis of the linguistic changes that have occurred from the acceleration of discourse and answering the question: What changes are produced, linguistically speaking, in millennial speech and writing?

Following Andrés Neuman, and his first micronote "The brief is not the same as the short: the brief shuts up in time, the short before time" (Neuman, 2012), we find ourselves before the emergence of a new way of making literature that is characterized by the micro space and linguistic brevity: tweeting.

In the words of Cecilia Colón, "twitter is a new way of writing very short stories based on a new technology that has influenced people who have access to a computer or any mobile device. Its format of only 140 characters distinguishes it and is a challenge for those who use Twitter as a means of communication" (Ortuño Lizarán, 2020). That is to say, the speech is shortened to the extreme, and the resource of ellipsis is constantly used in order to capture the reader and catch him in a thread.

Considering the second micropoint, "The vocation of every micro-story is to grow without being seen", tweeting is a way of making literature that, rather than grow in readers, does so in followers, which means that it has a growth in terms of to his media expectation. To give an example, one of the most famous tweeters is Spanish Manuel Bartual, who, in 2017, during his vacations, created a thread on Twitter with a micro-story of the horror and suspense genre, adding ingenuity games, thus, in less than one week it gained more than 300,000 followers, that is, 300,000 readers who longed for more stories of this type, serving as inspiration for future tweet writers, the purest microtextuality: see the writer Andrea Menéndez Faya (@MenendezFaya), Modesto García (@modesto_garcia) and Nagore Suárez (@NagoreSuarez), three of the most famous authors on Twitter in Spain today.

Twitter is committed to a stylistic change that also influences the reading of the text. This implies an acceleration in internal and external reading, with hardly any pauses, due to the absence of paragraphs and the shortening of words, something that complies with Andrés Neuman's third micronote: "The most particular thing about the micro-story is not its tiny length, but its radical structure" (Neuman, 2012).

There is also a commitment to freedom in the use of accentuation rules, that is, the content is more important than the form. Therefore, Neuman's fourth and fifth micronotes are fulfilled: the micro-story must be "Punctuated with a scalpel" and "A micro-story begins in quotation marks and ends in an ellipsis" (Neuman, 2012). In this sense, Nagore Suárez, winner of the public award in the special edition #YoMeQuedoEnCasa of the 'twitter' contest of the #FeriaDelHilo (@FeriaDelHilo): emphasizes the importance of knowing how to take advantage of "new formats and new times" and not to confuse the supports: "We are writing for Twitter" and you must "adapt the story to the format, not copy and paste something that we already have written" (Ortuño Lizarán, 2020). This non-normative use of linguistic and orthographic norms not only affects micronarrative but also micropoetry. In the words of Daniel Escandell Montiel and his article "Twitter: the frontier of microliterature in the digital space":

> The *tuitpoetry* is located in a line of tradition of brevity that places it in the sphere of haiku from a strictly technical perspective. Several authors publish micropoems that can be published in the space of a tweet, both in the form of poetic prose (to the extent that seen on the web, messages on Twitter do not accept full stop, but they do in specifc clients), or by resorting to to the bar to separate several verses, as in the case of @MicroPoesia: "the tongue of the afternoon/the lips of the night/the ardor of the dawn/the calm of the morning".
>
> (Escandell Montiel, 2014, p. 43)

It should be added that in these new discursive spaces, the separation of the author from his characters is also diffused, since the characters, on many occasions, are the authors themselves, since they tell autobiographical stories that require the interaction of the reader. Therefore, the complicity of the micro-story is needed through interaction and the indirect joke that supposes the challenge of reading the author with the character's clothing or vice versa. Neuman already specified it in his eighth micronote: "The characters of a microstory walk in profile" (Neuman, 2012). Therefore, and continuing with the ninth, "the micro-story needs brave readers, that is, those who support the incomplete" (ibídem), the incomplete in relation to the completion of the story in a short textual space, the trust of the character and the game. and the joke of the words themselves. As the tweet writer Nagore Suárez expresses: "My stories play with the advantage of being told live and in a way that sometimes makes readers doubt whether what they are reading is real or not" (Ortuño Lizarán, 2020). So, in part, this reading courage is also linked to the uncertainty of mixing fiction with reality, as a kind of unreliable narrator.

Finally, we end with the tenth point, which in itself involves the very reception of the micro-story, that is, the communicative pact without which the text could not have a significant place: "The shorter it seems, the slower it is read" (Neuman, 2012). Indeed, the brevity of the speech implies the search for strategies that favor the reading of the short story. Referring to the study *Strategies for reading in the micro-story*:

> As the short story constitutes an open fable (Eco, 1981, as cited in Neuman, 2012), the reader must carry out the process of conclusion or closure of the story, the only way to be able to construct a possible meaning. The inference process is built from the different conjectures and predictions that are offered about the continuity of history, to later develop the most coherent interpretations.
>
> (Larrea O, 2004)

This implies that the reader must be competent. It is not an easy literature but a strategic game that configures an interpretative pact between the author and the reader. We are, therefore, before a new identity expression that requires pacts of interpretation to find meaning. Pacts in discourse that are incomplete, like hybrid identities that metaphorize new ways of being in a world immensely full of unconnected halves, of hyperlinks of origin that lead to a new origin: life as a link to something else; that is how it is projected: a part of contemporary literature for being the channel of expression of a fleeting, accelerated, brief, digital voice and united by what it fragments.

3.4 CONCLUSION

Summarizing, we find ourselves before a new branch of contemporary literature that values, reads, studies and/or writes fragmented speeches, virtualized in social networks, fleeting at intervals of sometimes twenty-four hours before they disappear from the networks; interacting, accomplices, with their readers until their disappearance through comments and enhancing them through likes. The cybertext is no longer just a hypertext; now it becomes an option of a questionnaire so that the story continues towards one narrative line or another. Now the reader of the networks, such as Twitter or Instagram, is the protagonist of the plot twist, encompassed in a percentage that will have power in the choice of content.

But minification and its microtextuality are not exempt from media enemies and undervaluations, which is why writers of national relevance in the millennial generation are taking shape and delving into criticism, essential for the objective maturity of the texts for media dissemination. We speak, among other personalities, of Spanish writers such as Luna Miguel, who stated the following regarding this topic:

> To say that a writer with 20,000 followers on the social network of the day must necessarily be bad – and that if the publishing industry is interested it is only because of the number of potential clients among their likes – is almost like daring to say that a writer with a weekly column in the current newspaper is hopelessly bad – because if the publishing industry is interested, it is only because of the number of potential clients among the readers of such medium. In truth, both things could turn out to be more than true, with the exception that we will accuse the first of killing literature and the second we will crown him in the "best of the year" list.
>
> (Fernández, 2021)

In short, paradoxes that always haunt the breakaway movements with the already known forms. We are facing a linguistic change that involves the language and the literature that is born from it in new ways of making themselves known and reaching new readers. And that phenomenon always, before being criticized, is appreciated.

NOTES

1 The term was coined by Canadian sociologist Marshall McLuhan. The concept appears several times in his books *The Gutenberg Galaxy: The Making of Typographic Man* (1962) and *Understanding Media* (1964). In 1968, McLuhan used it in the title of his book *War and Peace in the Global Village*.

2 Term that appeared for the first time in his literary work *Homo Deus: Brief history of tomorrow*.

REFERENCES

Escandell Montiel, D. (2014). Tuiteratura: la frontera de la microliteratura en el espacio digital. *Iberic@l, 5*, 37–48.

Fernández, L. (2021). Las letras escondidas tras el filtro de Instagram. *El País*. https://elpais.com/cultura/2021-02-07/las-letras-escondidas-tras-el-filtro-de-instagram.html

Kim, S. (2022). *The metaverse: The digital earth-the world of rising trends*. Editorial Anaya Multimedia.

Lanzuela Corella, M. L. (1998). La literatura como fuente histórica. In B. P. Galdós (Ed.), *Actas del XIII Congreso de la Asociación Internacional de Hispanistas/ coord. por Florencio Sevilla Arroyo*, Vol. 2, 2000 (pp. 259–266). Carlos Alvar Ezquerra.

Larrea O, M. I. (2004). Estrategias de lectura en el microrrelato. *Estudio Filol. [en línea], 39*, 179–190. http://dx.doi.org/10.4067/S0071-17132004003900011.

Monterroso, A. (1959). *Obras completas (y otros cuentos)*. Editorial Anagrama.

Neuman, A. (2012). 10 micronotes on the short story. *Microrréplicas, Blog de Andrés Neuman*. http://andresneuman.blogspot.com/2012/12/10-microapuntes-sobre micronarrativa.html

Ortuño Lizarán, M. (2020). Tuiteratura (relatos): el arte de contar historias en hilos de Twitter. *El Obrero, Periodismo Transerval*. https://elobrero.es/cultura/47413-tuiteratura-el-arte-de-contar-historias-en-hilos-de-twitter.html

Sébart, O. (2016). La minificción: ¿el género literario del siglo XXI? *La Clé des Langues, Lyon, ENS de LYON/DGESCO*. https://cle.ens-lyon.fr/espagnol/litterature/litterature-latino-americaine/auteurs-contemporains/la-minificcion-el-genero-literario-del-siglo-xxi-

Sendra Ramos, S. (2018). *Elogio de lo mínimo. Estudios sobre microrrelato y minificción en el siglo XXI*. Iberoamericana/Vervuert.

Zavala, L. (2018). Elements for the purpose of minifiction precepts. *Microtextualidades: Revista internacional de microrrelato y minificción, 4*, 117–127.

CHAPTER 4

Metaverse and New Narrative

Storyliving in the Age of Metaverse

Dr. Kemal Deniz

Munzur University, Turkey

CONTENTS

4.1 INTRODUCTION

Recent technological advancements have facilitated several innovations and changes in the communication industry as well as other fields. Rapid developments in digital technologies reveal significant effects and results in this change and transformation. The effects of digital transformation, which can be observed in all areas, continue to create changes in a wide range, from daily life practices to communication channels and environments. Digital transformation has also caused significant changes in the interaction of the masses with communication tools and environments.

The era of analog mass media, dominated by a one-way communication process, has been digitally transformed by converged media. This current media occupies much more space in people's daily life practices, thanks to the advancements in telecommunication technologies.

The developments in telecommunication technologies, which started with Internet 2.0, especially highlighting user interaction with e-mails, blogs, and social networks, were the first stages of the impact of this digital transformation on large masses. With the invention of smart devices and improvements in mobile internet technologies, communication has become faster and more instantaneous. One does not have to be dependent on a stationary place and wired connections anymore. People had the opportunity to stay connected to the network at anytime and anywhere without interruption. Thus, they had an alternative way to interact more on social, cultural, and personal levels. These developments and technological jumps, which emerged in communication environments and were both caused and accelerated by digital transformation, continue to reveal innovations that will change the social, cultural, and economic relations that affect people's daily lives.

According to many social media or new media researchers, this process has also had a strengthening effect on participatory democracy (Wilhelm, 2000; Coleman & Blumler, 2007; Loader & Mercea, 2011), even though others have some reasonable concerns (Persily & Tucker, 2020). It contributed to a period that radically transformed the centralized and corporate communication processes and questioned the reliability of news and other corporate communication methods used in mass media. New communication tools that function as alternative channels to traditional media continue from citizen journalism (Allan, 2013; Blaagaard, 2018; Wall, 2019; Nah & Chung, 2020) increasing with social media to sharing video-activism actions (Harding, 2001; Gregory et al., 2005; Ristovska & Price, 2018; Ristovska, 2021) thanks to images and videos that can be transmitted instantly. The spread of these interactions continues to be influential in the emergence of results that contribute to the public's ability to gain rights against authority. This atmosphere created by digital transformation is still an efficient mechanism and hope for independent and rapidly spreading alternative ideas to be expressed globally to build up common sense by reaching large masses.

The most basic virtual interaction is also experienced through these social networks since the interaction that individuals have via Internet 2.0 technology is most commonly on social networks. According to the research in January 2023, 4.76 billion were social media users out of 5.16 billion internet users worldwide (Statista, 2023). However, this process continues to move to a different dimension through VR technologies. Thus, individuals who can stay connected in the virtual environment in the context of interaction possibilities previously limited by social networks continue to be a part of a communication process that brings more visuality to the fore with new technologies. This process allows particularly visual communication to be carried to different dimensions.

VR technology continues to be influential by bringing novelty to various applications, from playing digital games (Swink, 2009) to cultural heritage visualization and museum applications (Parry, 2010). By using virtual reality glasses and other aiding devices, VR technology also supports education from primary school to university level (Daniela, 2020), and to simulate actual experience for areas that require professional training in serious games (SG) (Cai et al., 2021; Cai & Cao, 2021). Movies and serials broadcast on online platforms have become interactively structured (McErlean, 2018). Unlike the traditional story structure, the audience has not only become a passive spectator of the narratives but also has the role of being an active participant in the process that interferes with the course of the story and finalizes different endings by choosing various options.

The most advanced level of these technologies will probably be the metaverse, one of the most current developments in today's communication technologies. The process, which started with the opening of the door of the virtual world with VR, seems to reach a new dimension with the introduction of the metaverse concept. Because the metaverse concept has quickly entered our lives as both a new communication medium and a phenomenon. In addition to the innovations and effects it will bring to all other economic and financial processes such as crypto coin, NFT, blockchain, and Web3, the metaverse will also reveal a new narrative paradigm.

Thus, how the narratives will be shaped in this new communication and interaction environment requires generating theoretical insights not only technologically but also in the context of social, cultural, and communication research. It would be notable to consider the narrative form in the metaverse era, which can be seen as a rise of innovations from storytelling to the interactive narrative that emerged with digital transformation. It is also significant to consider the large inclusiveness of the meta-narrative for the whole, the specificity of the narrative about each subject or object in the metaverse, and how this will be reflected in the big picture. In the metaverse era, the inquiries about how the transition of the narrative to the *storyliving* stage instead of storytelling will affect or conclude the ontology of the processes of creating the narrative, transmitting it through a medium, and experiencing it should be considered. This study as an argumentation on the concept of the metaverse questions the narrative in terms of the social, cultural, and economic dynamics that could be transformed by the metaverse.

4.2 META

Meta emerged into our lives as a new concept rapidly. Large masses became informed about it in a short time as a phenomenon called the metaverse. From technology gurus to digital art enthusiasts and communication scientists, a wide range of professionals has started to research and follow developments about the metaverse idea with attention. Before Facebook, which is one of the most established networks in social media platforms, took the corporate name "Meta", the CEO of the company, Mark Zuckerberg, announced the metaverse, his model of metaverse (Newton, 2021), which aroused great interest and curiosity.

Metaverse is a combination of the words meta and universe. Therefore, the metaverse conceptually expresses the notion of a metaverse. The word meta (Merriam-Webster, n.d.) is an adjective in the self-referential sense, and it is a word that changes, transforms, or gives a meaning to express the beyond and a transcendent level if it is used with another word or used as a prefix. Metaverse was described by Ball (2022) as

> a massively scaled and interoperable network of real-time rendered 3D virtual worlds that can be experienced synchronously and persistently by an effectively unlimited number of users with an individual sense of presence, and with continuity of data, such as identity, history, entitlements, objects, communications, and payments.

Cyber meta-reality is defined as "the infinite multiverse of realities that can be experienced, inhabited, created, and shared by humans" (Sipper, 2022, p. 1), that have the inclusive expression of the metaverse conception. "After the Metaverse, the distinction between virtual and non-virtual reality will come to an end or will shift to a different perspective, so the reality of the Metaverse will gain meaning and value as a reality on its own (Deniz, 2022, p. 96)." For a new entity in a digitally created environment, metaverse will probably include or combine all the digital reality and immersive technologies

(Doerner et al., 2022) such as virtual reality (VR), augmented reality (AR), mixed reality (MR), and human-computer interaction (HCI) under the umbrella of extended reality (XR).

The genuine and unique product will be licensed to its owner with a non-fungible token (NFT), enabling them to do everything from creating an avatar for themselves in the metaverse era to purchasing homes and land in cities that will take place in the metaverse. A new ownership and organization paradigm for the virtual environment will probably become increasingly common as a result of this development. Obviously, this commodity has a financial value that can be converted into actual money on cryptocurrency exchanges. In the metaverse, you will be able to sell and swap your investments, including everything from digital art to real estate, from cryptocurrency to conventional currency or financial instruments (Davis, 2021; Gonzales, 2021; Russel, 2021; Clemens, 2022; Stock, 2022). It is an indication that the nature of international trade is about to change in the age of metaverse.

It has been possible to put forward various definitions and predictions on the concept of metaverse since the earliest academic studies (Zagalo et al., 2011). However, examples can be given from creative productions and artworks made in the past that can be a reference for us in establishing a framework for the concept of metaverse. There are productions created both in the field of literature and in audio-visual media such as movies and TV series that can enlighten us about what kind of environment the metaverse could be in. We can develop a perspective that will help us embody this environment, in the context of their stories and narratives and the universe imagination they offer us visually.

The name metaverse was first mentioned in Neal Stephenson's fantasy science fiction novel *Snow Crash* in 1992. Metaverse, in which the character of the novel, Hero, takes place, is like the real world but does not exist physically. As a software created with a computer protocol and presented to the public via the internet network, the metaverse is expressed as follows: "He's [Hero] in a computer-generated universe that his computer is drawing onto his goggles and pumping into his earphones. In the lingo, this imaginary place is known as the Metaverse (Stephenson, 1992)." In *Matrix* (1999), we learn from Morpheus as a dystopian way that

the Matrix is everywhere. It is all around us. Now in this very room. You can see it when you look out the window or when you turn on the television. You can feel it when you go to work . . . when

you go to church . . . when you pay your taxes. . . . It is the World
that has been pulled over your eyes to blind you from the truth. . . .
Unfortunately, no one can be told what the Matrix is. You have to
see it for yourself.

Among the movies that have the metaverse concept, *Matrix*'s (1999) futuristic reality proposition differs from the digital game universe in the fiction
of *Ready Player One* (2018), which was adapted from the 2011 written science fiction novel by Ernst Cline with the same name (Cline, 2011). Today,
the technological development required for the existence of the metaverse
environment has emerged without the need for fantasy or science fiction.
While today's technology has come to a level that will make all people
swallow the red pill in *The Matrix*, the metaverse as a concept seems to
continue to evolve to become an alternative technology and interactive
environment to the physical world, beyond being an irreversible parallel
universe of our lives.

Black Mirror was broadcast in 22 episodes between 2011–2019 and 5
series on Netflix, and an interactive movie was added to the series in
2018. The characters in these series are presented in the form of a meta-
self or being in the context of time, space, or their own existence in the
environment they are in (Black Mirror, 2019). Episodes of the series often
conclude with dystopian and unhappy endings. In addition to these
series, *Altered Carbon* (2020) was adapted from Richard K. Morgan's
cyberpunk novel of the same title, whose narrative takes place in a kind
of metaverse (Morgan, 2002). We can accept these fictional narratives as
examples of a metaverse set in virtual environments. However, the lives
of real individuals in the metaverse experience and their interaction in
the physical, mental, and ontological context of this environment would
be different from a fictional hero who can acquire and use the superpow-
ers provided by technology.

One can consider the metaverse as the simulation of real entities or
includes the simulations of those entities belonging to the real world. From
a simulation perspective, it is not Baudrillard's simulacra thoughts but
more likely close to his hyperreality concept (1994) that offers the blend-
ing nature of real and artificial so that one cannot be distinguished the
difference. In a metaverse, the real assets should be copied into the representative models as digital twins in an immersive environment. The digital
twins might be not only small objects, pieces of stuff, or some living entities but also cities, countries, or planets as huge complex compositions.

That is the meta-self-experience of the immersive hyperreality will probably blend more than ever in the metaverse before it was in the actual world's digital or virtual interactions.

While we can make some accurate predictions about what the metaverse is, there are still many unanswered questions. In the upcoming years, it could become more evident what its virtual environment will look like, how it will impact people's social and cultural lives, and how it will affect the actual world's economy. Due to the potential of Web 2.0, we are still talking about people who can create a virtual-self in a virtual environment and develop a social and cultural identity with a set of values unique to the digital world.

4.3 META-SELF

As the research in the field of artificial intelligence and machine technology develop, it is seen that there are transformed cyborg-like persons whose bodily or mental functions have improved in applying treatment by providing more comfortable living conditions in medical fields (Miah & Rich, 2008; Le et al., 2018). We can also see these characters and beings in a narrative environment. The fictional heroes of classical science fiction and futuristic narratives, in which computer technologies have transformed and added superior powers or new functions, can be seen as character types that readers and viewers from literature to cinema are highly familiar with.

The meta-self, which will be more like reality TV characters with a specific goal, or a framed universe limited to today's TV program format, will be the subject of the narrative, like a game character. Thus, it can be predicted that the experience of the real person watching the VR images created in the digital environment will be a different interaction process from the experience of the meta-human and her/his meta-self who experiences the metaverse in the context of the narrative and shares its ontology. In this sense, it is also necessary to evaluate the digital age shaped by digital environments and technologies, which is still the most influential audiovisual age globally, and how the transformation of the virtual-self, which are the individuals of these environments, can be (Slater & Sanchez-Vives, 2014). It would be very appropriate to assess the narrative in the context of the metaverse, which is predicted to start another era in the future, both at the levels of communication in general and social and cultural interaction concerning the new kind of person, her/his virtual character, and meta-self.

Every moment we are in the virtual environment, we reconstruct ourselves according to the structural, social, and cultural characteristics of this virtual environment. In other words, we have a unique self in this virtual world, and it continues to evolve when we are in this environment. Thus, this virtual-self which is built in the virtual environment has more existential significance for us when we are online. Even in the real and offline world, we shape our identity, personality, and behavioural patterns with the structural and cultural characteristics of virtual environments (Papacharissi, 2002). The digital transformation, which has already taken place today, offers us new codes and rules in our social interactions, beyond the old traditional communication methods of daily life. Therefore, in these highly digitalized interaction environments, the selves of individuals become and transform into this environment in which they maintain their existence to a large extent and rebuild themselves. In this case, it can be observed that the most basic essence of our existence is our presence in the digital environment, our usages here, and the traces we leave – in other words, our data in the virtual space that represents us.

In addition to virtual environments accessed through social networks and interactive usage practices, which are now very ordinary and common today, research aiming to overcome the physical limits of the human brain continues, especially thanks to neuroscience, medical, and advanced computer technologies (Clément, 2019; Opris et al., 2021). BCI (brain-computer interface) technology aims to connect an unlimited cloud environment to the brains of individuals so that the neural-nanorobotics-enhanced human brain can instantly access all the information in the accessible virtual environment. These studies will be about access in various usage areas between education to entertainment (Martins et al., 2019). Thus, people will be able to process unlimited amounts of data while accessing it. This will enable people to integrate into the digital environment much more and use this environment as an extension of their brains.

Under the shadow of the concerns about these transformations (Vidal, 2022), the technological, medical, and ethical functioning of this process will also carry the human being to a kind of post-human or meta-human level (Haney, 2006). If this happens, the human mind and, accordingly, the self will have reached a different level from its previous practices with its existence in the virtual environment beyond its physical existence and reality. The human being, who can have the experience of the virtual environment through her/his senses, by using or wearing virtual reality devices, through the integration of the brain and computer, she/he will now be able to create this virtual world in her/his mind. This would be a new experience about the process that receives and transmits the virtual environment or any kind of data to a person's brain with a helmet-like device with some linkages to the neural receptors. For example, when a person reaches this universe by establishing a neurotechnological connection in a metaverse whose existence is virtually complete, she/he will no longer be an audience of it but a resident of that environment. Therefore, she/he will experience the metaverse as a meta-human with her/his meta-self. This meta-self, however, could easily be created in the form of any kind of avatar or other entities such as we saw the meta-selves in another body of different sex, age, skin colour, and nationality in *Altered Carbon* (2020).

The virtual-self and the reconstruction of this virtual-self to a higher level of reality in the metaverse can also be explained with a post-human or meta-human concept. Thus, in the context of the metaverse phenomenon, the concept of the post-interactive virtual reality narrative in the immersive environment should be discussed with the conceptualization of *storyliving* (Deniz, 2022, p. 97). Storyliving will be experienced by the human being as a meta-self who will take it to a different process from the storytelling stage of the traditional dimension. Meta-self will be, then, turned into a part of the narrative in the virtual environment with her/his virtual-self through metaverse and meta-human interaction.

4.4 STORYLIVING IN THE METAVERSE

In the metaverse, the set and plot may be similar or very different from the stories created by the real-world time-space and interactions that take place there. These will offer depending on the variations in the context of the development of metaverse technology and the integration of humans into it, and perhaps our efforts to make sense under new categories and classifications will continue. However, today we can evaluate the narrative in the metaverse concerning its own ontology as a virtual environment in terms of structure, character, and set so that we can make further predictions in this context.

4.4.1 Structure

The plot, time, and space, that is, setting, character, and narrator, which are the most basic elements in traditional storytelling, generally constitute the basic components of the narrative (Aristotle, 2006). The plot is the element that covers what the narrative is basically, the sequence of events that the narrative draws attention to. The setting includes the space and time in which the plot emerges and develops. A character is an individual who is the subject or participant in the plot. The narrator, on the other hand, may be the author, the director, the main character of the narrative or another character or group of characters in this narrative.

These principal elements of the classical narrative are seen in examples ranging from the first mythological stories to epics, from novels to motion pictures, and episodic series of television dramas. Therefore, to tell us the elements of a place and time in which the events shaped around or directing a character take place has reached from the stories that people gathered around the fire told and listened to each other before the written culture. Then it continued from starting with the printing press to the dramatic or reality narratives produced by audio-visual tools and finally has been created in different environments by computer and digital devices for VR applications and games (Aylett & Louchart, 2003; Kress, 2020) that have been experienced by the audience.

This process maintains from the most primitive to the most advanced technology supported. In short, although the media and the technologies used have changed, the classical storytelling and structural set-up have not changed much. In this context, the adventures of the characters continued to show similarities and maintain the familiar narrative structure (Propp, 1968; Campbell, 2004), even though the time, place, and subject changed.

Texts that transform the classical narrative structure with a modern style continue to try to create fiction by following different methods such as not conveying the beginning, middle, and end of the story in a linear time and story construction order. However, beyond this classical narrative approach, some experimentation opportunities started after the innovations brought by technology increased the production possibilities and options for structuring the narrative. First, video and then computer games, namely digital games, began to offer multi-choice story structures in which the player, as the protagonist of the event, directs the narrative or plot with active participation. The player had the opportunity to have this experience interactively.

In multi-character and multiplayer games, through the interaction of different characters, the players sometimes allowed unwritten events or situations or storylines to occur in the game (Louchart & Aylett, 2004; Walsh, 2011). Especially, these encounters in-game plots or the fact that the player does not follow the planned game plot and engages in a different fictional or personal interaction has recently been a subject of interest to narrative scholars as an example of "emerging narrative". In the metaverse environment, emergent narrative can be expressed as the intersection of the individual or private metaverse experiences of different characters in a common or collaborative environment and the spontaneous storyliving experience that emerges accordingly. Thus, the interaction of each subject's own storyliving adventure with someone else's narrative (plot) will also create an intertextuality or an intertextual match.

In an interactive structure that we met more recently, a narrative structure has emerged in which the course of the story changes as a result of the decisions we make and the choices we did. The interactive narrative is sometimes constructed like a game and leaves the selection of certain turning points in the plot to the person, who are in the position of the audience, participant, and player. The first examples of interactive narrative can be seen in Jorge Luis Borges' 1941 written story *The Garden of Forking Paths* (El jardín de senderos que se bifurcan), and novels and adventure stories for children such as *Choose Your Own Adventure* published by Bantam Books between 1979 and 1998. Interactive narrative is also known as hypertext fiction or interactive fiction (Douglas, 2000).

It is possible to experience examples of interactive drama in audio-visual environments, such as in the interactive category of Netflix where movies and animations are listed (Netflix, n.d.). Amongst them, *Black Mirror: Bandersnatch* (2018) is an example of an interactive narrative in the form of a feature-length fiction film. These interactive narratives require the viewers and the participants in the fiction to make decisions about the options they encounter and thus make rational choices to reach the end.

4.4.2 Character

The individual will become both the character and the inhabitant of the narrative in the metaverse. In other words, in the context of this narrative, the self will become a part of the virtual reality of this environment, and the virtual-self will have to undergo separation and transformation ontologically from the real world's self as the subject of the narrative.

While the virtual-self stands out in the context of social networks and interactive narrative by being reconstructed by digitalization, the meta-self would be the person living in the metaverse as an element of the post-virtual narrative. In this context, whether the narrative in the metaverse is a temporal and spatial copy of the real world, or it is fictionalized with a completely different space-time and plot, the person who experiences the metaverse will be the real person who participates in this narrative with her/his meta-self by storyliving through it. Therefore, individuals will become the residents of the narrative regardless of the story or plot in the metaverse.

Thus, we can consider the self of the individual that has been digitized in the virtual reality of the metaverse and her/his existential being adapted to this virtual environment just like a post-virtual narrative version of the micro-universes in which today's reality narratives are created for television in the real world.

The ontology of the character and how the meta-self, which has become a part of the virtual universe, will undergo a change, metamorphosis, or existential transformation might be illuminated in more detail by the impressions obtained from the experiences of individuals in the future. The meta-self should be investigated from the perspective of humanities scholars such as sociologists, social psychologists, and researchers studying digital visual culture and transformations in the interactional processes of storyliving in the metaverse environment. Moreover, if neuroscientists became clearer in understanding how the human brain and its mental functioning can or cannot adapt to this interaction with the metaverse and the processes experienced, that would be a great contribution to the post-virtual narrative.

4.4.3 Virtual Setting

The metaverse might be summed up as a further dimension of keeping traditional social, cultural, and financial values in a virtual setting. A person can reside in this metaverse with her/his own virtual avatar, giving her/him a chance to forge an identity and sense of self. They may outfit their avatars, feed them, and give them a home or piece of land, cars, or other objects, etc. by purchasing these properties like they would do in the actual world. Real estate as well as mobile property can be purchased as digital possessions using crypto coins and NFTs. Thus, persons will be able to create a new existence represented by their avatars in the metaverse with the virtual environment digitally equipped with worldly stuff and entities.

Some cities, like Dubai, Singapore, Seoul, and Shanghai, are already being mentioned as potential metaverse participants through the creation of their digital twins (Cohen, 2022). Either creating some fictitious places to inhabit or just copying a part of the actual world as a virtual representation for conducting daily practices on the metaverse will be shaped as a substituted dimension for a person's virtual existence that interacts to create and develop her/his identity as a meta-self.

For the people with fully digitalized virtual identities who experience metaverse, they need to inhabit, change, and recreate the world around them like they do in the actual world. That means the struggle to construct both the virtual reality and their virtual selves must be experienced interactively as a storyline that narratives need. The narrative that is set in a virtual environment performed by a person or the virtual-self of this person is the process of storyliving. That is the virtual setting such as the metaverse is a medium for storyliving.

4.5 CONCLUSION

The concept of metaverse, which we began hearing frequently with the metaverse phenomenon, seems to continue to be mentioned together with some other technological developments associated with it. In particular, the metaverse on a worldwide scale and the opportunities it might bring will continue to be discussed in the context of its effects at the economic level.

In the context of the future of communication, metaverse, using VR within a vast shared digital network environment beyond the limits of today's internet technology that is simply a way to eliminate the time and space limitations of all the interactions we perform in our offline world, refers to a new era that we can experience through our virtual presence in this environment or our avatar representing this existence. Thus, we can foresee that we will be able to perform all our interactions on a more virtual level in this new communication era.

When we evaluate the condition of individuals in the context of metaverse and their position in this environment, we can assume that their virtual selves, which have come to the fore in social media and other interactive online environments in recent years, may become more and more compatible with the ontology of the metaverse. Thus, it can be expected that individuals in the metaverse will take place in a new narrative environment, and all their interactions there will be aligned with a meta-self that is unique to that place, like a kind of reality TV character. Watching digital content interactively with VR technology will be advanced to the level of post-virtual narrative in the age of metaverse. When individuals in the metaverse complete the digital transformation of their meta-selves socially, culturally, and behaviourally, they will become a part of the narrative and experience storyliving in their interactions.

Consequently, virtual reality, meta-self, and metaverse environment as the dynamics in the interactive processes of living a story in digital culture will remain the main issues waiting to be examined with an inter-disciplinary approach in the coming years. It will also be an essential contribution to the field of the post-virtual narrative if it becomes more understandable how human beings can adapt to the metaverse and the interaction in this environment without losing their human values that have been shaped by the actual world and without losing the existence that this world has brought to them through very long evolutionary processes. The ontology of the character and how the meta-self, which has become a part of the virtual universe, will undergo a change or existential transformation can be illuminated in more detail in the future with the impressions obtained from the storyliving experiences of individuals in the metaverse.

REFERENCES

Allan, S. (2013). *Citizen witnessing: Revisioning journalism in times of crisis.* Polity Press.

Aristotle. (2006). *Poetics* (Trans. J. Sachs). Focus Publishing.

Aylett, R., & Louchart, S. (2003). Towards a narrative theory of virtual reality. *Virtual Reality, 7,* 2–9. https://doi.org/10.1007/s10055-003-0114-9

Ball, M. (2022). *The metaverse: And how it will revolutionize everything.* Liveright Publishing Corporation.

Baudrillard, J. (1994). *Simulacra and simulation* (Trans. S. F. Glaser). University of Michigan Press.

Blaagaard, B. B. (2018). *Citizen journalism as conceptual practice: Postcolonial archives and embodied political acts of new media.* Rowman & Littlefield.

Cai, Y., & Cao, Q. (2021). *When VR serious games meet special needs education research, development and their applications.* Springer.

Cai, Y., Van Joolingen, W., & Veermans, K. (2021). *Virtual and augmented reality, simulation and serious games for education.* Springer.

Campbell, J. (2004). *The hero with a thousand of faces.* Princeton University Press.

Clemens, A. (2022). *Metaverse for beginners: A guide to help you learn about metaverse, virtual reality and investing in NFTs.* Independently Published.

Clément, C. (2019). *Brain-computer interface technologies: Accelerating neuro technology for human benefit.* Springer International Publishing.

Cline, E. (2011). *Ready player one.* Crown Publishers.

Cohen, B. (2022). The 4 metaverse cities to watch beyond 2022. *The Smart City Journal.* Retrieved November 21, 2022, from www.thesmartcityjournal.com/en/cities/the-4-metaverse-cities-to-watch-beyond-2022

Coleman, S., & Blumler, J. G. (2007). *The internet and democratic citizenship: Theory, practice and policy.* Cambridge University Press.

Daniela, L. (2020). *New perspectives on virtual and augmented reality: Finding new ways to teach in a transformed learning environment.* Routledge.

Davis, W. J. (2021). *Metaverse explained for beginners a complete guide to investing in cryptocurrency, NFT, blockchain, digital assets, web 3 & future technologies.* Independently Published.

Deniz, K. (2022). Metaverse: Real data in the virtual universe. In N. Çokluk & N. Kara (Eds.), *Privacy in the digital age* (pp. 81–107). Literaturk Academia.

Doerner, R., Broll, W., Grimm, P., & Jung, B. (2022). *Virtual and augmented reality (VR/AR): Foundations and methods of extended realities (XR).* Springer.

Douglas, J. Y. (2000). *The end of books – Or books without end? Reading interactive narratives.* The University of Michigan Press.

Gonzales, D. (2021). *Metaverse investing: How NFTs, web 3.0, virtual land, and virtual reality are going to change the world as we know it.* Independently Published.

Gregory, S., Caldwell, G., Avni, R., & Harding, T. (2005). *Video for change: A guide for advocacy and activism.* Pluto Press.

Haney, W. S. (2006). *Cyberculture, cyborgs and science fiction consciousness and the posthuman.* Rodopi.

Harding, T. (2001). *The video activist handbook* (2nd ed.). Pluto Press.

Jones, A., & Brooker, C. (Creators/Executive producers). (2019). *Black Mirror* [TV series]. Channel 4/Netflix. https://www.netflix.com/

Kalogridis, L. (Creator/Executive producer). (2020). *Altered Carbon* [TV series]. Netflix. https://www.netflix.com/

Kress, B. C. (2020). *Optical architectures for augmented-, virtual-, and mixed-reality headsets.* Spie Press.

Le, D. N., Le, V. C., Tromp, J. G., & Nguyen, G. N. (2018). *Emerging technologies for health and medicine: Virtual reality, augmented reality, artificial intelligence, internet of things, robotics, industry 4.0.* Scrivener Publishing, Wiley.

Loader, B. D., & Mercea, D. (2011). Networking democracy? Social media innovations in participatory politics. *Information, Communication and Society, 14*(6), 757–769. https://doi.org/10.1080/1369118X.2011.592648

Louchart, S., & Aylett, R. (2004). Narrative theory and emergent interactive narrative. *International Journal of Continuing Engineering Education and Lifelong Learning, 14*(6), 506–518.

Martins. N. R. B., Angelica, A., Chakravarthy, K., Svidinenko, Y., Boehm, F. J., Opris, I., Lebedev, M. A., Swan, M., Garan, S. A., Rosenfeld, J. V., Hogg, T., & Freitas Jr, R. A. (2019). Human brain/cloud interface. *Frontiers in Neuroscience, 13*, 112. https://doi.org/10.3389/fnins.2019.00112

McErlean, K. (2018). *Interactive narratives and transmedia storytelling: Creating immersive stories across new media platforms.* Routledge.

Merriam-Webster. (n.d.). Meta. *Merriam-Webster.com Dictionary.* Retrieved December 17, 2022, from www.merriam-webster.com/dictionary/meta

Miah, A., & Rich, E. (2008). *The medicalization of cyberspace.* Routledge.

Morgan, R. K. (2002). *Altered carbon.* Gollancz.

Nah, S., & Chung, D. S. (2020). *Understanding citizen journalism as civic participation.* Routledge.

Netflix. (n.d.). Interactive. *Netflix.com*. Retrieved December 19, 2022, from www. netflix.com/search?q=interactive

Newton, C. (2021). Mark on the Metaverse: Facebook's CEO on why the social network is becoming 'a metaverse company'. *The Verge*. Retrieved December 11, 2022, from www.theverge.com/22588022/mark-zuckerberg-facebook-ceo-metaverse-interview

Opris, I., Noga, B. R., Lebedev, M. A., & Casanova, M. F. (2021). Modern approaches to augmentation of brain function: Brain-computer interfaces. In I. Opris, M. A. Lebedev, & M. F. Casanova (Eds.), *Modern approaches to augmentation of brain function: Brain-computer interfaces* (pp. 57–89). Springer Nature. https://doi.org/10.1007/978-3-030-54564-2_4

Papacharissi, Z. (2002). The presentation of self in virtual life: Characteristics of personal home pages. *Journalism and Mass Communication Quarterly*, *79*(3), 683–660. https://doi.org/10.1177/107769900207900307

Parry, R. (2010). *Museums in a digital age*. Routledge.

Persily, N., & Tucker, J. A. (2020). *Social media and democracy: The state of the field, prospects for reform*. The Social Science Research Council.

Propp, V. (1968). *Morphology of the folktale*. University of Texas Press.

Ristovska, S. (2021). *Seeing human rights: Video activism as a proxy profession*. The MIT Press.

Ristovska, S., & Price, M. (Eds.) (2018). *Visual imagery and human rights practice*. Palgrave Macmillan.

Russel, J. (2021). *Metaverse for beginners: A complete guide on how to invest in the metaverse*. Independently Published.

Sipper, J.A. (2022). *The cyber meta-reality: Beyond the metaverse*. Lexington Books.

Slade, D. (Director). (2018). *Black Mirror: Bandersnatch* [Film]. Netflix. https://www.netflix.com/

Slater, M., & Sanchez-Vives, M. V. (2014). Transcending the self in immersive virtual reality. *Computer*, *47*(7), 24–30. https://doi.org/10.1109/MC.2014.198.

Spielberg, S. (Director). (2018). *Ready Player One* [Film]. Warner Bros.

Statista. (2023). Worldwide digital population January 2023. *Number of internet and social media users worldwide as of January 2023*. Retrieved May 11, 2023, from www.statista.com/statistics/617136/digital-population-worldwide/

Stephenson, N. (1992). *Snow crash*. Spectra/Bantam.

Stock, B. (2022). *Metaverse: The #1 guide to conquer the blockchain world and invest in virtual lands, NFT (crypto art), altcoins and cryptocurrency + best DeFi projects*. Blockchain NFT Academy.

Swink, S. (2009). *Game feel: A game designer's guide to virtual sensation*. Morgan Kaufmann Publishers.

Vidal, C. (2022). Neurotechnologies under the eye of bioethics. *eNeuro*. *9*(3). https://doi.org/10.1523/ENEURO.0072-22.2022

Wachowski, L., & Wachowski, L. (Directors). (1999). *The Matrix* [Film]. Warner Bros.

Wall, M. (2019). *Citizen journalism practices, propaganda, pedagogy*. Routledge.

Walsh, R. (2011). Emergent narrative in interactive media. *Narrative, 19*(1), 72–85. https://doi.org/10.1353/nar.2011.0006.

Wilhelm, A. G. (2000). *Democracy in the digital age: Challenges to political life in cyberspace*. Routledge.

Zagalo, N., Morgado, L., & Boa-Ventura, A. (2011). *Virtual worlds and metaverse platforms: New communication and identity paradigms*. Information Science Reference.

CHAPTER 5

Customised Metaverse

The Study of Factors Influencing the Use of Humanlike Avatars of Social Media Content Creators in Communication with Their Audience

Dr. Grzegorz Ptaszek
AGH University of Krakow, Poland, and EduVR Lab AGH

Dr. Damian Gałuszka
AGH University of Krakow, Poland, and EduVR lab AGH

Dr. Jowita Guja
AGH University of Krakow, Poland, and EduVR Lab AGH

Dr. Tomasz Masłyk
AGH University of Krakow, Poland, and EduVR Lab AGH

CONTENTS

DOI: 10.1201/9781003379119-5

5.1 INTRODUCTION

Avatars, defined as "digital entities with anthropomorphic appearance, controlled by a human or software, that have an ability to interact" (Miao et al., 2022), are increasingly present in mediated media communication, ranging from social media to online shopping assistants, medical applications and communication between remote collaborators (Nowak & Fox, 2018). Digital humans are also appearing as virtual influencers on social media (Instagram, TikTok, YouTube) and are used in avatar marketing as a strategy to support brands (de Brito Silva et al., 2022). This is meant to be a way to capture the attention of audiences in a market saturated with social media influencers, who are becoming less authentic in promoting brands (Awdziej et al., 2022).

With the announcement of a new dimension of the metaverse – communication taking place in a VR environment – advocates believe that the use of digital avatars as new modes of communication in a global virtual reality environment will be attractive enough to drive social media users to become more connected. The avatar will act as an alter ego of real people in key spaces of its user's life – work, entertainment, social and family interactions – transferred to virtual reality.

Marketing research show that the new dimension of the metaverse is attractive primarily to the younger generation of internet users (18–24), mainly men (YouGov, 2022). While entertainment and leisure activities most interest UK consumers, those in the US gravitate more to socialising with friends/family. The situation is similar for Polish consumers, for whom, according to a survey of 700 people, the most attractive aspects of the metaverse are the opportunities to meet with friends (83%) and to meet new people (81%) as well as the possibility to look the way they want (79%) and to be who they want (78%).[1]

At the same time, however, the metaverse is a huge space with commercial potential, a fact that the companies involved in its development are very well aware of: manufacturers of VR/AR hardware and applications, owners of popular social media, developers of platforms using blockchains technology, game designers making money from micropayments and finally companies opening virtual outposts (e.g. BMW's Joytopia) on platforms such as Blocktopia, Next Earth or Decentraland anticipating their rapid growth in popularity. Such platforms function in a similar way to the much earlier *Second Life*, but in addition to a computer or a phone, they are or will eventually be accessible from VR or AR goggles. They also assume a much more extensive level of monetisation. If these predictions come true, this will also translate into the popularity of real-life influencers in the metaverse and the problem of the digital body will become significantly more important.

The research discussed in this chapter aims to confront these processes with the public's willingness to participate in them. They are distinguished by the fact that they focus on the perspective of a specific group: people profiting from their own image in the digital space, influential creators publishing their content on channels such as YouTube or TikTok. The anticipated 'metaverse-isation' of media can be both a source of profit for them and a challenge they will not be able to overcome.

5.2 LITERATURE REVIEW

The issue of avatar-mediated communication and its attractiveness has been addressed in numerous research publications to date. Among others, researchers have been interested in the effectiveness of communication via 3D avatars viewed on screen (Mull et al., 2015) and in a VR environment (Wu et al., 2021; Garau et al., 2003), or in the evaluation of avatar attractiveness and the factors shaping it (Mull et al., 2015; Pakanen et al., 2022; Blom et al., 2014; Bailenson & Segovia, 2010). These studies suggest, inter alia, that behavioural realism plays a more important role in communication through avatars (not necessarily anthropomorphic ones) than visual realism, for example, accurate representation of facial expressions as well as the ability to control gaze direction, which, in the case of realistic avatars, has a positive impact on evaluating the level of face-to-face interaction, co-presence and interlocutor evaluation (Garau et al., 2003).

In turn, research into the attractiveness of avatars examined issues such as the level of realism (Garau et al., 2003), the similarity of the avatar to the subject (Blom et al., 2014; Bailenson & Segovia, 2010), the type of avatars (human, fantasy, animal, humanoid, etc.) (Mull et al., 2015), context (e.g., gaming, social media, sales) (Mull et al., 2015; Pakanen et al., 2022; Trepte & Reinecke, 2010), full-body avatar versus avatars represented only by a body or bust (Pakanen et al., 2022). These studies show that, in general, human avatars (Mull et al., 2015), photorealistic full-body-type avatars (Pakanen et al., 2022), subjectively perceived by the subject as similar to him or her (Blom et al., 2014; Bailenson & Segovia, 2010), are rated highest in terms of attractiveness. The importance of the context and environment in which an avatar is used has also been demonstrated. For example, in a study of perceptions of avatars as virtual assistants in online shops, a fantasy avatar was rated just as highly as a human avatar (Mull et al., 2015), while in computer games, the photorealism of the avatar did not determine its attractiveness (Pakanen et al., 2022). However, as Trepte and Reinecke's (2010) show, the role of avatars in computer games is more complex, because it can depend on a type of a game, personality and even the player's perceived satisfaction with life. This research also found that players preferred avatars that resembled them in terms of personality in non-competitive games and when they themselves were satisfied with their lives. In contrast, they chose dissimilar avatars in competitive-based games when they were less satisfied with their lives.

Hitherto, only limited research on avatars have investigated the role of personality factors. An example is the use of the Big Five theory to determine the relationship between respondents' personality and the analysed avatar's traits, which, in practice and using extensive questionnaires (e.g., BFI-44, BFI-10), has been determined in the dimensions of neuroticism, extraversion, conscientiousness, openness to experience and agreeableness (Bélisle & Bodur, 2010; Trepte & Reinecke, 2010; Fong & Mar, 2015).

The findings discussed previously are highly relevant to our study, because the avatars that we examine are ultimately to be used in a specific communication and commercial context, characterised by behavioural realism (they accurately represent facial expressions) and resemblance to the user. However, the context in which they are to be used is very specific, since they are to be used by streamers and other video content creators hitherto monetising their own physical image. This seems relevant to the results of our study, especially if one considers the relationship described by Garau et al. (2003), that as the realism of an avatar increases, so do the expectations of its realistic behaviour. Despite this, avatars can act as digital assistants to support online communication and e-commerce (Blunk et al., 2020; Lee et al., 2015).

It is worth mentioning that, depending on a type of an avatar, it may generate different reactions in potential customers. Avatars reproducing the image of a professional (e.g., a dermatologist on a cosmetics manufacturer's website) are more likely to arouse cognitive trust, whereas attractive avatars to arouse emotional trust – both of which may influence the willingness to purchase a particular product (Lee et al., 2015). Similarly, the impact of avatars in educational contexts has also been problematised. Research by Ratan et al. (2022) found that, despite the expected influence of the Proteus effect – according to which people tend to tune their behaviour with the identity characteristics of their avatars – students' use of avatars representing their ideal-self or future-self, rather than their actual-self, during remote learning was sequentially associated with slightly lower exam grades or slightly lower exam grades and lower self-efficacy.

5.3 MATERIALS AND METHODS

The study used research methods and techniques typical for 3D avatar research. So far, researchers have used, among other things, questionnaires, or evaluation surveys (depending on the design of a particular study), filled in by groups of several to several dozen respondents before or after an exposure to the avatar. The exposure itself was usually implemented by showing the avatar in a photo, video or, alternatively, through the participation of respondents in a virtual environment, if the technology used allowed it (e.g., a VR application, a digital game, or a virtual world like Second Life) (Bélisle & Bodur, 2010; Mull et al., 2015; Pakanen et al., 2022). Like the present study, 3D avatar research has also utilised distance and online research methods using widely available services such as Discord, Skype or Google Hangouts (Pakanen et al., 2022), as well as 3D avatars reflecting the appearance of real people (Frampton-Clerk & Oyekoya, 2022).

5.3.1 Participants

Participants (N=64) were influencers and streamers, that is, people involved professionally or occasionally in creating and publishing online video content on weblogs or social media (e.g. Facebook, Instagram, TikTok, Twitch). Recruitment for the study was done in a targeted manner. The invitation to participate in the study was sent out individually to video creators working with influencer marketing agencies: LTTM – Central Europe's largest group providing this type of service – and Tears of Joy. Respondents received free temporary access to ExplodedView's advanced avatar-based video creation platform, AvatarCam, in exchange for taking part in the study. Twenty-six women and 36 men took part in it; 2 people did not wish to specify their gender.

5.3.2 Stimulus Material

The video was produced in collaboration with ExplodedView using the AvatarCam smartphone app. The application creates a humanlike 3D avatar based on a previously taken selfie and then, using a camera installed in a smartphone or a laptop that tracks human facial movements in real time, animates the avatar's behaviour in a virtual environment. Three characters were prepared in the application: two male and one female, whose behaviours were presented to the subjects in a short video (180 sec). The three characters differ in appearance, clothing, behaviour and occur in two different virtual environments (outer space and office) (Figure 5.1). In order

FIGURE 5.1 Images of the physical characters (top) and their humanlike avatars (bottom) used in the study.

to be able to compare the appearance and behaviour of the 3D humanlike avatar with the appearance and behaviour of the real character animating the avatar, the avatar's performance was introduced by a 5-second statement from the real character (e.g., "Hi, my name is Marcin, and I want to show you my world without borders").

5.3.3 Measurements Instruments

The study measured variables such as human-avatar evaluation, self-esteem, body-esteem – appearance, the sense of efficiency, and personality traits.

5.3.3.1 Human – Avatar Scale

This variable was measured using the author's tool, consisting of 5 contrasting pairs of statements relating to both people and the avatars representing them (more attractive – less attractive, more friendly – less friendly, stronger in attracting attention – less in attracting attention, less intimidating – more intimidating, facilitates understanding the communication – hinders understanding the communication), rated on a 7-point Likert scale (1 – clear advantage of the avatar over the real person, and 7 – clear advantage of the real person over the avatar).

5.3.3.2 Self-Esteem

This variable was measured using the free standardised tool Rosenberg Self-Esteem Scale (Rosenberg, 1989) in the Polish adaptation (Łaguna et al., 2007), consisting of a 10-item Likert-type scale with a 4-point metric (1 – strongly agree; 4 – strongly disagree). The scale is a one-dimensional tool allowing for the assessment of the level of general self-esteem – a relatively stable disposition understood as a conscious attitude (positive or negative) towards the self. The Polish version of the method is a reliable tool (& Cronbach's 0.81–0.83), with confirmed theoretical accuracy.

5.3.3.3 Body-Esteem – Appearance

To measure this variable, the Body-Esteem – Appearance subscale of the Body-Esteem Scale for Adolescents and Adults (BESAA) by Mendelson et al. (2001) in a Polish adaptation (Słowińska, 2019) was used. This subscale contains 10 items rated by the respondent on a 5-degree Likert scale (1 – never, 5 – always), relating to general feelings about their appearance (e.g., 'I like what I see when I look in the mirror', 'I like how I come out in pictures'). The subscale has satisfactory reliability for the adult population (Cronbach's α 0.90).

5.3.3.4 The Sense of Efficiency

We measured this variable using a standardised, free-of-charge tool, the Sense of Efficiency Test (Chomczyńska-Rubacha & Rubacha, 2013), a 17-item Likert-type-scale with a 4-point metric (1 – definitely yes, 4 – definitely no). The tool measures self-efficacy understood as an individual's personal resources shaping its conviction about the effectiveness of taken actions such as ability to defer gratification, ability to translate goals into an action program, self-confidence, resistance to frustration and stress, persistence in action, sense of agency, developmental motivation, inner directedness. The items of the scale are arranged into two groups of resources conditioning self-efficacy: cognitive-activity (8 items) and motivational (9 items). The overall reliability index is Cronbach's α for adults of 0.87, where individual items ranged from (0.86–0.89). However, only the motivational group of resources conditioning self-efficacy (9 items) was used in the analysis.

5.3.3.5 Personality Traits

To measure personality traits in the Big Five model (neuroticism, extraversion, conscientiousness, openness to experience and agreeableness), we used the Ten Item Personality Inventory (TIPI) by Gosling et al. (2003) in a Polish adaptation (Sorokowska et al., 2014). The 10-statement inventory comprised of the five scales (corresponding to the 'Big Five' personality factors), starting with the words: I perceive myself as a person. The subject is asked to respond to each self-description on a 7-point Likert scale (from 1 – strongly disagree to 7 – strongly agree). All the TIPI-PL scales have similar or higher reliability to the subscales of the original version and are, depending on how the questionnaire was completed (paper – pencil, Google sheet, web application) (Cronbach's α): Extraversion – 0.68–0.74, Agreeableness – 0.50–0.58, Conscientiousness – 075–0.80, Emotional stability – 0.65–0.83, Openness to experience – 0.44–0.47.

5.3.3.6 Procedure

The research was conducted online between July–November 2022. The invitation to participate in the study was sent directly to online video content creators with a link to the questionnaire preceded by the instructions and information about data confidentiality and personal data processing. The tool was created using Microsoft Forms in the university version. First, the respondents were asked to watch a 180-second video (see part Stimulus material) displaying the possibilities offered by the use of an avatar created from a photo of a physical person. They then answered questions grouped into several sections: evaluation of the person-avatar (5 items), overall evaluation (1 item), anticipated audience reactions (1 item), propensity to use the avatar (2 items), characteristics of activities undertaken online (6 items), Rosenberg Self-Esteem Scale (10 items), the Sense of Efficiency Test (17 items), Body-Esteem of Appearance of the BESAA (10 items), TIPI (10 items).

5.4 RESULTS

The aim of the study was to verify the assumption that the humanlike assessment of a 3D avatar depends on the impression it makes on the study participants when confronted with physical persons who create and animate their avatar (human-avatar scale), controlling the influence of personality variables: self-esteem, body-assessment – appearance, self-efficacy, and personality traits. This assumption was verified using a linear regression model in which the dependent variable was the avatar score, and the independent variable was the score on the human-avatar scale.

Participants evaluated the avatar based on three questions relating to:

a) the overall impression ("The avatar shown in the presented video can be used for various online activities. What overall impression did it make on you?");

b) the anticipated evaluation of the audience ("Whatever your evaluation of this avatar is, please assess how your audience would react if you used this avatar").

c) willingness to use ("Given the online activity you do, assess whether you would use this avatar for your own purposes").

All three questions were based on 5-point scales, the extremes of which were described respectively: "very bad – very good", "definitely negative – definitely positive" and "definitely would not use – would definitely use – would use". The survey participants' responses to these three questions were characterised by high internal consistency (Cronbach's α 0.785), and therefore an 'avatar rating' index was created based on them. Its values were the arithmetic mean of the values obtained on each question.

The independent variable described the impression (positive-negative) that the avatar had on the study participants when confronted with the physical individuals who used it (*human – avatar assessment*). The favorable opinion of the avatar increased as the value on the scale increased. Ratings of the individual aspects proved to be internally consistent (Cronbach's α 0.748) and were therefore linked on a single scale. The assessment indicator was the arithmetic mean of the scores obtained on the individual scales. As such, it was included in the regression model as an independent variable.

Personality variables were treated as control variables in the model. It was assumed that study participants could 'step into the shoes' of the avatar user and evaluate the avatar as a tool to hide a multitude of vices or highlight advantages. All variables had decent reliability, respectively: self-esteem (Cronbach's α 0.871), body-assessment – appearance (Cronbach's α 0.867), the sense of efficiency (motivational resources) (Cronbach's α 0.765), personality, including extraversion (Cronbach's α 0, 605), agreeableness (Cronbach's α 0.506), conscientiousness (Cronbach's α 0.724), emotional stability (Cronbach's α 0.678) and openness to experience (Cronbach's α 0.547).

5.4.1 Avatar Assessment: Descriptive Statistics

Respondents (N=64) rated the avatar "neither good nor bad" (mean 2.9). It should be stressed, however, that the rather positive impression after the presentation of the avatar (mean score of 3.3) was not matched by an assessment of the anticipated reaction of potential recipients of messages in which such an avatar could be used (mean score of 2.7). Perhaps this was the reason for the general, rather reluctant attitude of the survey participants towards the use of an avatar as part of their own internet activity (mean 2.8).

According to the respondents, the confrontation between avatar and human came out more favourable for the avatar (mean 4.2). Most notably, in favour of avatars, was that they were less intimidating (4.8), attracted more attention (4.3) and seemed friendlier (4.2). Aspects such as attractiveness and understanding the message were rated slightly lower (3.9) (see Table 5.1).

5.4.2 Avatar Assessment: Dependency Model

The assumption that the overall assessment of an avatar depends on the impression it makes on observers when confronted with physical individuals was verified using a linear regression model, the results of which are shown in Table 5.2.

TABLE 5.1   Descriptive Statistics of Avatar Ratings and Personality Traits of Study Participants

Variable		Mean
Avatar rating (scale 1–5)	General assessment	2.9
	Impression	3.3
	Anticipated audience response	2.7
	Willingness to use	2.8
Human-avatar scale	General assessment	4.2
(scale 1–7)	Attractiveness	3.9
	Friendliness	4.2
	Attracting attention	4.3
	Overawing	4.8
	Understanding the message (communicativeness)	3.9
Self-esteem (scale 1–4)		3
Body-esteem - Appearance (scale 1–4)		2.8
The sense of efficiency (motivational resources) (scale 1–4)		2.7
Personality traits (scale 1–7)	Extraversion	5
	Agreeableness	5.4
	Conscientiousness	5
	Emotional stability	4.2
	Openness to experience	5.2

TABLE 5.2   Relationship between the Evaluation of an Avatar and the Impression It Gives When Confronted with a Real Person: Linear Regression Model Estimation Results

Variable	B	Std. Error.	Beta
Human-avatar scale	.389	.091	.529*
Self-esteem	.373	.284	.239
Body-esteem – Appearance	-.195	.253	-.116
The sense of efficiency (motivational resources)	-.366	.296	-.208
Extraversion	.017	.070	.029
Agreeableness	.003	.087	.004
Conscientiousness	.119	.068	.214
Emotional stability	.021	.080	.040
Openness to experience	.017	.099	.025
Constant			

(Continued)

TABLE 5.2   *(Continued)*

Variable	B	Std. Error.	Beta
		Model summary	
Adjusted R²		0.311	
Std. Error of the Estimate		0.76	
Test F		p<0.001	
N		64	

Source: Note: ˙dla p<0.01

The predictive value of the presented model is moderate (adjusted R2 = 0.311). The only significant predictor of the avatar's evaluation in this model found the impression of the avatar's presentation on the study participants in comparison with a human. The better the avatar was rated in this confrontation, the higher its overall rating was too. None of the personality traits was a significant predictor of avatar assessment.

5.5 DISCUSSION

The aim of the study was to verify the assumption that the humanlike assessment of a 3D avatar depends on the impression it makes on the study participants when confronted with physical persons who create and animate their avatar (human-avatar scale), while controlling the influence of personality variables: self-esteem, body-esteem – appearance, sense of efficacy, and personality traits.

The results of the study show that the evaluation of the avatar compared to the physical person who created it, based on its own image, fell more positively in favour of the avatar. The avatars appeared to be less intimidating, attracted greater attention of the respondents and seemed friendlier, equally rated as less attractive and less communicative. One explanation for this assessment may be that the 3D avatars, although created based on a photograph of a real character, were not characterised by a high degree of realism of form (like, for example, digital humans), but had features of cartoon avatars, such as unnaturally large eyes or a simplified representation. However, this somewhat paradoxical conclusion fits in with the findings of other researchers, who indicate that attempts to create humanlike avatars do not always lead to better results relative to anthropomorphic avatars, if only because of the Uncanny Valley effect (Jo et al., 2017).

A significant predictor of avatar evaluation was the impression that the avatar's presentation made on the participants of the study compared to a human, while the personality traits of the respondents were not a differentiating factor in this evaluation. Thus, it seems that the introduction of an avatar would make sense at the beginning of an influencer's activity when he or she is just creating his or her own image and can thus shape the audience's attitude to the message itself and to himself or herself as the sender (a hypothesis worth testing in future studies). The recipient will then judge his or her credibility, just like the virtual influencers, without comparison to the person representing the avatar.

It is worth highlighting that the positive assessment of avatars did not mix with their willingness to use them in their work. This shows that content creators are attached to their own image and fear that they may be perceived as inauthentic, which can only exacerbate viewers' uncertainty when interacting with an avatar. However, as the results of research on virtual influencers show, when it comes to credibility, it does not matter to the audience whether they know (or not) if the character they are interacting with is virtual or real (Awdziej et al., 2022). Given the results of our research, an avatar creator intending to use an avatar to promote its own video content should be aware of how his or her avatar modifies the actual image of the person. When an avatar is created with the intention of representing a specific person, its attractiveness and realism will be judged in relation to the real character. Considering the findings of previous research, not only visual realism (of form) but, above all, behavioural realism is therefore important. Although it should be noted that the representation of a person through a realistic humanoid avatar may increase the audience's expectations of realism (Garau et al., 2003). In other words, audiences of messages implemented through such an avatar, for example, fans of an online creator, may expect the avatar to also represent specific characteristics of the human prototype's appearance and behaviour. Depending on the specific implementation or creation of the avatar, as well as the level of technology behind it, these expectations may be fulfilled to a greater or lesser extent, influencing the overall feelings and evaluations towards the used avatar.

5.6 LIMITATION

In this study, the assessment of the avatar and the possibility of using it in the activity undertaken by online video content creators was based only on a comparison of the avatar capabilities presented in a 180-second video. It seems that a situation in which the respondents create their own avatar and test its capabilities could influence their evaluation. The video presented the subjects from interacting with the avatar and assuming a context other than the one imposed in the video. Nevertheless, the surveyed creators had the opportunity to evaluate the avatar from the perspective of their own viewers, which seems decisive. Indeed, it is the anticipated reactions of content viewers that become crucial to the choices made by creators of online video content.

Moreover, it should be presumed that the behaviour of the actors and the actress, including factors such as appearance, tone, manner or order of speech, may influence respondents' reception and evaluation of the 3D avatars. So can the contexts of use of the 3D avatars, which can also condition their reception. However, at a deeper level of analysis, important differences may already become apparent between avatar types (e.g. animated vs. photorealistic avatars), online platforms (e.g. YouTube vs. Twitch) or media technologies (e.g. video broadcasts vs. VR environments). Thus, the issues addressed in this chapter are evidently complex and require further research. We argue that future studies should construct an experimental environment in which the above-mentioned factors remain under the control of the researchers and their possible mediation in the evaluation of avatars is measured.

Acknowledgment

The authors would like to thank Marcin Klimek, Rafał Guzik and Jagoda Piluch from ExplodedView team for organizational support of this research. The research was funded by ExplodedView under a Research & Development agreement with AGH University of Krakow.

NOTE

1 https://nowymarketing.pl/a/38313,gameset-zbadal-potencjal-metaverse-marketingu-w-polsce) [21.12.2022].

REFERENCES

Awdziej, M., Plata-Alf, D., & Tkaczyk, J. (2022). Real or not, doesn't matter, as long as you are hot: Exploring the perceived credibility of an Instagram virtual influencer. In M. Awdziej & J. Tkaczyk (Ed.), *Extending boundaries: The impact of the digital world on consumers and marketing* (pp. 33–46). Kozminski University.

Bailenson, J. N., & Segovia, K. Y. (2010). Virtual doppelgangers: Psychological effects of avatars who ignore their owners. In W. S. Bainbridge (Ed.), *Online worlds: Convergence of the real and the virtual* (pp. 175–186). Springer.

Bélisle, J. F., & Bodur, H. O. (2010). Avatars as information: Perception of consumers based on their avatars in virtual worlds. *Psychology & Marketing, 27*(8), 741–765.

Blom, K. J., Bellido Rivas, A. I., Alvarez, X., Cetinaslan, O., Oliveira, B., Orvalho, V., & Slater, M. (2014). Achieving participant acceptance of their avatars. *Presence: Teleoperators and Virtual Environments, 23*(3), 287–299.

Blunk, O., Brown, G., Osmers, N., & Prilla, M. (2020). *Potentials of AR technology for the digitalization of consultancy intensive sales processes on the example of furniture sales.* Retrieved December 20, 2022, from https://doi.org/10.30844/wi_2020_t1-blunk

Chomczyńska-Rubacha, M., & Rubacha, K. (2013). Test Poczucia Skuteczności. Opracowanie teoretyczne i psychometryczne Pracowni Narzędzi Badawczych Komitetu Nauk Pedagogicznych PAN. *Przegląd Badań Edukacyjnych, 1*(16), 85–105. https://doi.org/10.12775/PBE.2013.007.

de Brito Silva, M. J., de Oliveira Ramos Delfino, L., Alves Cerqueira, K., & de Oliveira Campos, P. (2022). Avatar marketing: A study on the engagement and authenticity of virtual influencers on Instagram. *Social Network Analysis and Mining, 12,* 130. https://doi.org/10.1007/s13278-022-00966-w

Fong, K., & Mar, R. A. (2015). What does my avatar say about me? Inferring personality from avatars. *Personality and Social Psychology Bulletin, 41*(2), 237–249. https://doi.org/10.1177/0146167214562761

Frampton-Clerk, A., & Oyekoya, O. (2022). Investigating the perceived realism of the other user's look-alike avatars. In *28th ACM symposium on virtual reality software and technology* (pp. 1–5). Association for Computing Machinery. https://doi.org/10.1145/3562939.3565636

Garau, M., Slater, M., Vinayagamoorthy, V., Brogni, A., Steed, A., & Sasse, M. A. (2003, April). The impact of avatar realism and eye gaze control on perceived quality of communication in a shared immersive virtual environment. In *Proceedings of the SIGCHI conference on human factors in computing systems* (pp. 529–536). Association for Computing Machinery. https://doi.org/10.1145/642611.642703

Gosling, S. D., Rentfrow, P. J., & Swann, W. B. Jr. (2003). A very brief measure of the Big-Five personality domains. *Journal of Research in Personality, 37,* 504–528.

Jo, D., Kim, K., Welch, G. F., Jeon, W., Kim, Y., Kim, K. H., & Kim, G. J. (2017). The impact of avatar-owner visual similarity on body ownership in immersive virtual reality. In *Proceedings of the 23rd ACM symposium on virtual reality software and technology* (pp. 1–2). Association for Computing Machinery. https://doi.org/10.1145/3139131.3141214

Łaguna, M., Lachowicz-Tabaczek, K., & Dzwonkowska, I. (2007). Skala samooceny SES Morrisa Rosenberga – polska adaptacja metody. *Psychologia Społeczna, 2*(4), 164–176.

Lee, H., Sun, P., Chen, T., & Jhu, Y. (2015). The effects of avatar on trust and purchase intention of female online consumer: Consumer knowledge as a moderator. *International Journal of Electronic Commerce Studies, 6,* 99–118.

Mendelson, B. K., Mendelson, M. J., & White, D. R. (2001). Body esteem scale for adolescents and adults. *Journal of Personality Assessment, 76*, 90–106.

Miao, F., Kozlenkova, I. V., Wang, H., Xie, T., & Palmatier, R. W. (2022). An emerging theory of avatar marketing. *Journal of Marketing, 86*(1), 67–90. https://doi.org/10.1177/0022242921996646

Mull, I., Wyss, J., Moon, E., & Lee, S.-E. (2015), An exploratory study of using 3D avatars as online salespeople: The effect of avatar type on credibility, homophily, attractiveness and intention to interact. *Journal of Fashion Marketing and Management, 19*(2), 154–168. https://doi.org/10.1108/JFMM-05-2014-0033

Nowak, K. L., & Fox, J. (2018). Avatars and computer-mediated communication: A review of the definitions, uses, and effects of digital representations. *Review of Communication Research, 6,* 30–53. https://doi.org/10.12840/issn.2255-4165.2018.06.01.015

Pakanen, M., Alavesa, P., van Berkel, N., Koskela, T., & Ojala, T. (2022). "Nice to see you virtually": Thoughtful design and evaluation of virtual avatar of the other user in AR and VR based telexistence systems. *Entertainment Computing, 40,* 100457. https://doi.org/10.1016/j.entcom.2021.100457

Ratan, R., Klein, M. S., Ucha, C. R., & Cherchiglia, L. L. (2022). Avatar customization orientation and undergraduate-course outcomes: Actual-self avatars are better than ideal-self and future-self avatars. *Computers & Education, 191,* 104643. https://doi.org/10.1016/j.compedu.2022.104643

Rosenberg, M. (1989). *Society and adolescent self-image* (Revised ed.). Wesleyan University Press.

Słowińska, A. (2019). Skala Samooceny Ciała dla Adolescentów i Dorosłych (BESAA) – polska adaptacja metody. *Polish Journal of Applied Psychology, 17*(1), 21–31.

Sorokowska, A., Słowińska A., Zbieg A., & Sorokowski, P. (2014). *Polska adaptacja testu Ten Item Personality Inventory (TIPI) – TIPI-PL – wersja standardowa i internetowa.* WrocLab.

Trepte, S., & Reinecke, L. (2010). Avatar creation and video game enjoyment: Effects of life-satisfaction, game competitiveness, and identification with the avatar. *Journal of Media Psychology: Theories, Methods, and Applications, 22*(4), 171–184. https://doi.org/10.1027/1864-1105/a000022

Wu, Y., Wang, Y., Jung, S., Hoermann, S., & Lindeman, R. W. (2021). Using a fully expressive avatar to collaborate in virtual reality: Evaluation of task performance, presence, and attraction. *Frontiers in Virtual Reality, 2,* 641296. https://doi.org/10.3389/frvir.2021.641296

YouGov (2022). *Unlocking the metaverse. An analysis into UK & US attitudes about the buzzy new technology.* Retrieved January 2, 2013, from www.iab.com/wp-content/uploads/2022/10/Metaverse-8.22.pdf

CHAPTER 6

Metaverse and Diversity

Dr. Erkan Saka

Istanbul Bilgi University, Turkey

CONTENTS

6.1 INTRODUCTION

Diversity may be defined as the variety of differences between people in a group or organisation, such as different backgrounds, cultures, genders, abilities, experiences, and perspectives.[1] The term has become essential in theorising business or social interactions because of its supposed nature to promote creativity, innovation, and collaboration. It is also believed to help create a more equitable workplace and society by providing opportunities for people from different backgrounds and perspectives to contribute. Additionally, it can help increase customer engagement, as diverse

DOI: 10.1201/9781003379119-6

73

organisations can better understand and respond to the needs of their customers (Dobbin & Kalev, 2013; Ely & Thomas, 2020). Thus, diversity and related and supportive terms such as equity, equality, and inclusion[2] are frequent items on the corporate agendas and success recipes (Putriani & Aras, 2022; Ramirez, 2021). Companies regularly point out their successes in this field through press releases "Infosys was also recognised [in addition to women workforce] for inclusion of persons with disabilities, LGBTQIA+ and other strands of diversity, making it an inclusive workplace for all (Infosys, 2022)." Diversity in the advertising and marketing industries has reached an all-time high, according to data from the Association of National Advertisers (ANA) (Ruelas, 2022). Social media platforms such as Tiktok and Twitter announced their explicit goals to reach a higher level of diversity (Putriani & Aras, 2022).

A recent report (Averstad et al., 2022) on the state of diversity in global private markets, with a particular focus on private equity (PE) firms, like in many other fields of business, demonstrates that diversity is poor despite the recent emphasis on DEI. The report primarily focused on gender and ethnic or racial diversity within PE firms. The presence of minorities in PE investing teams tends to be lower in higher-ranking positions, and there are more obstacles to women's career advancement. Despite the rising consciousness of gender equality, the metaverse industry is not different in inequalities. According to Estrada (2022), the metaverse lacks women in leadership roles: Members of Open Metaverse Alliance for Web3 and Metaverse Standards Forum (Responsible Metaverse Alliance, 2022) are mostly led by men; 8% and 9% of them have Female CEOs, respectively. Men-led metaverse initiatives received $107 billion in funding compared to those led by women, which received $5 billion. In the meantime, women who are active in the metaverse are more likely to be power users than men – 35% compared to 29% (Estrada, 2022).

As Hollowell (2022) states, the lack of diversity in the tech space has direct social consequences. From algorithms to AI-based systems and metaverse, there is growing literature on the outcomes of various sorts of bias. Erken (2022), for instance, points out that users with disabilities will be affected by metaverse usage. The author frames the issue regarding socio-digital inequalities and elaborates how even the types of disabilities may lead to other inequalities (Dobransky & Hargittai, 2016). In the current metaverse building, visually impaired citizens are more likely to be exposed to usage disadvantages than types of disabled citizens. Erken argues that due to the difficulties in the design and cost of virtual reality technologies, there is a potential to exclude specific segments of society. In this respect, few scholarly studies focused on integrating the disabled into the virtual reality experience (i.e. Zhao et al., 2018 and more examples in Erken, 2022).

This study is not an exhaustive review of diversity issues in the tech space. However, in an exploratory research fashion (Stebbins, 2001), the author reviewed secondary sources that focused on metaverse, diversity, and related concepts. The term metaverse took a new turn with the rebranding of Facebook as Meta in 2021, and the author tracked the emergence of metaverse's reemergence since then. This chapter outlines the emerging issues in this context and aims to connect these to the ongoing critical literature on bias issues in technology platforms.

A direct connection to diversity debates in metaverse is the existence of bias. From algorithm studies to big data and artificial intelligence studies, a rising concern is the bias issues (from gender to racial or cultural perspectives) embedded in the core technical infrastructures. In most cases, corporations recognise bias and promise to make fundamental changes. The European Union is already making policies to challenge the possible adverse outcomes of bias in Big Tech (Engler, 2022). A recurring theme in the public debates is the lack of diversity among the developers and decision-makers in infrastructural personnel, in the big data used, and in software implementations. Metaverse appeared as a core concept as part of a cluster of other terms such as Web3 and blockchain. Many businesses and brands aim to participate in an emerging metaverse scene, and there is an appealing journalistic interest in the topic.

What I want to do is to focus on how issues of diversity engage with the current building process of metaverse. As the metaverse is growing through interrelated technologies such as virtual and augmented reality, as well as new developments in artificial intelligence and blockchain applications, current debates and imaginaries are important because they may shape the building process we live through.

6.2 PROBLEM OF BIAS IN BIG TECH

In their critique of algorithms and the quest to understand bias, Shorey and Howard (2016) point out that many big data projects fail to consider information ethics adequately. Specifically, they argue that these projects often lack consideration for the impact on individuals and society and tend to simplify and strip away the complexity and context of social systems. Additionally, these projects often rely on personal information to create large data sets and use methods of analysis that do not account for preexisting biases. Historical conditions (Barnes & Wilson, 2014; Dalton, 2013), control and production of big data (Thatcher, 2014), the subjects of big data (Haklay, 2013), and application contexts of big data (Kitchin & Dodge, 2011) are some of the prominent factors in data creation. Thus, data does not actually "transcend context or domain-specific knowledge" (Kitchin, 2014, p. 4).

Biased data may then be integrated and disembodied into the platform. As digital media occupy a more central place in the economy and society, major tech corporations are standard setters that include biased algorithms. Google's "power through the algorithm" (Beer, 2009) is already an old but powerful argument. Mager's (2018) study demonstrates how Google is believed to be setting the standards de facto internationally, from data protection to many other privacy-related issues. Carah and Angus (2018, p. 3) extend this to other significant players by referring to Van Dijck (2013): "Platforms like Facebook and Instagram create interfaces, protocols, databases and algorithms that engineer and optimise user participation for commercial-ends." Going back to the data issue, Eisenstat, the elections integrity operations head at Facebook from June to November 2018, claimed that many AI and machine learning systems are being trained with what she calls "bad data." Tech companies have mandatory "managing bias" training to help with diversity and inclusion issues. However, these ignore the field of cognitive bias and decision-making (Eisenstat, 2019). My study on gender bias (Saka, 2020) demonstrates that in addition to calling for data set diversification, there must be more diversity among algorithm designers and builders and transparency in the development processes to mitigate bias. In the following section, concerns about the diversity among the builders of metaverse occupy a central place. However, most ongoing debates are carried out in this new stage with varying degrees.

6.3 ISSUES OF DIVERSITY IN METAVERSE

6.3.1 Bias in AI

A direct source of the lack of diversity may be tied to bias in artificial intelligence (AI). The bias discussions move from big data and algorithms to AI. A well-noted example is Amazon's discontinued AI-based tool to recruit employees. Based on the data for the past ten years, the tool had completely ignored women candidates and recommended male ones (Parihar, 2022). Parihar argues that a fair AI will produce a more objective recruitment process and ensure an equitable employee training system. This will lead to real diversity and inclusiveness in the workplace. To achieve this, a diversity of workforce at the outset may be essential. Parihar refers to a recent study that found less than 6 percent of Google employees were Latin, and only 3.3 percent were Black. She also emphasised that the representation of women in AI or even tech teams overall is very low. As an example remedy, she praises Microsoft, which set up a "Fairness, Accountability, Transparency, and Ethics in AI" team to uncover any biases that can be found in data used by its AI systems.

Spatial aspects of metaverse connect it to the debates of AI usage in building smart cities. Brandusescu and Reia (2022) aim for AI usage to create more civic engagement and, in the meantime, overview the inherent risks. Here Datta (2022) believes that the metaverse can be used to build smart, resilient, and inclusive cities in the future. AI tools for the cities can utilise the physical data from its many sensors and build information models, digital infrastructure, and geospatial information to replicate and create models in the metaverse. These can be used for seniors with mobility issues and physically disabled residents to offer them accessible services and opportunities to visit businesses and attend events.

6.3.2 Users, Generations, Stakeholders

Some ideas on diversity are tied to the new user base, generation changes, and a new assemblage of stakeholders. The hyped expectations of the Generation Z (users who were born between 1995 and 2010) are visible in some of them. A press statement states that Gen Z is known for valuing diversity, inclusive culture, justice, and equity (Press Release Network, 2022), and they may create a fairer and more inclusive world (Bullfrag Blog, 2022). Here, platforms like Meta will provide to build our own worlds, and it will be these users' responsibility to generate content that will be safe and free from discrimination and harassment (Dupuy, 2022). Another observation about the metaverse user base is that women spend more time in the metaverse than men. Women are more likely than men to engage in hybrid use cases in the metaverse (that is, using both physical and digital worlds to take part in activities such as gaming, fitness, education, live events, and shopping via AR/VR technologies) (Alaghband & Yee, 2022). As will be noted in the next section, these positive statistics about female users do not convert to leadership roles and do not significantly help the state of diversity in the workforce.

To its credit, Dupuy was quick to distribute responsibility to other stakeholders: No single entity can build an equitable and inclusive metaverse, and there must be cooperation between the private sector, legislatures, civil society, academia, and users. New forms of collaboration between developers and government officials are needed and must be imagined to go beyond the existing real-world inequalities and not reproduce current problems in digital communications (Dupuy, 2022; *Fast Company*, 2022). Diversity among creators seems to be the link to starting collaborative examples. In a piece at *Fast Company* (2022), brands are urged to involve creators not at the last minute but from the outset so that meaningful and substantive products can emerge. Rhonda X and LGBQT+ creator Nathan "Skitter" Crawford to design Mastercard's True Self World is given as a successful case (*Fast Company*, 2022).

On the other hand, some scholars attribute decentralised architecture a more significant role (Lacity et al., 2022). Users will be empowered with privacy, security, equity, and inclusion in a decentralised metaverse. Still, the lack of economic incentives and business models for decentralised applications could reproduce current internet ownership patterns and dystopian qualities such as negative consequences of addiction, cyberbullying, surveillance, and cybersecurity breaches, among others. However, there is not a single decentralisation process (Shardeum, 2022; Zarrin et al., 2021), and further research should be devoted to the potential of decentralisation.

6.3.3 Citizens, Not Customers

Diversity may not be uttered, but there are strong calls to make the metaverse citizen-centric. Hyper-commercialisation of the current internet is behind these calls. While Peyton (2022) uses a manifesto style to make the metaverse accessible to all citizens, some others point out the hyper-capitalism in the metaverse projects: Instead of a digital universe to free human creativity, everything is tied to crypto wallets that are powered by NFTs which, in turn, are oriented towards transactions to be logged in a blockchain (Digital Trends, 2022). Horwood (2022) quotes Amadeep Sirha on the importance of focusing on humans rather than technology. In doing this, metaverse may be safer than the traditionally male-dominated physical places. Metaverse may be inclusive not only for women but also for neurodiverse people (Horwood, 2022). Sun et al. (2022) approach from a very different angle: The use of big data analytics in the creation of prediction models and decision-making processes is projected to grow dramatically as people shift away from avatar keyboards in immersive virtual environments. Humans will become data-producing entities to unprecedented degrees. This is a critical point in terms of privacy and will be noted in another section. Still, the emphasis is that there could be more accessible data to build models for a more diverse human population. However, this may not be achieved before preventing significant early diversity issues.

6.3.4 Lack of Women in Leadership and Reproduction of Gendered Roles

A recurrent theme is the state of women in metaverse. McClain (2022) presents a bleak picture when she discusses how real-world beauty standards are maintained in some virtual communities. Many beauty insecurities are carried into the metaverse as many women avatars with wide hips, ample busts, and tiny waists are visible. Women users are exposed to more harassment and derogatory statements.

Women seem to be locked out of the leadership roles in the metaverse businesses. Alaghband and Yee (2022) believe this is critical for creating and setting metaverse standards in its early stages. Ninety per cent of leadership roles are held by men in several standards bodies that aim to set interoperability norms for the metaverse (i.e. the Metaverse Standards Forum and the Open Metaverse Alliance for Web3 (OMA3). The authors quote McKinsey's June 2022 report on the state of metaverse, where the gender gap in leadership positions is too big. The authors argue that this is not different from the overall business world patterns (less than 10 per cent of Fortune 500 CEOs are women). As male technology pioneers sink billions in unsuccessful metaverse projects, the gender hierarchy is preserved (Consultancy.uk, 2022).

These findings are in line with the broader tech world workforce. Despite calls for a more diverse cyber workforce, some developed countries such as Japan, Germany, and the US have the least diversity rate, including the gender gap (Leonard, 2022).[3] "Diversity deficit" is also noted in Sharma (2022), who points out an Instahyre report of the 25:75 ratio tech talent market. Getting more women into science, technology, engineering, and mathematics (STEM) is still a challenge (Sharma, 2022), and there is even a decline in women employees in the tech industry (Schijns, 2022).

Ryder (2022) also starts by pointing out more statistics about the gender gap: Only 5 per cent of crypto companies are estimated to be led by women, and 19 per cent of people who own digital currency identify themselves as women. According to a report by The Female Quotient and EWG Unlimited and quoted by Ryder, 75 per cent of women have heard of the metaverse, but only 30 per cent claim that they are truly familiar with it.

However, despite the gender disparity, Ryder informs the readers about the initiatives to promote women: While Qualia[4] is an incubator for start-ups, BFF[5] acts as an educational community to educate women on crypto. NFT collective World of Women[6] partners with the fashion industry to encourage more female-identifying individuals to explore Web3. Another case is from China, Web3 Women Union,[7] which aims to bring more women into Web3 in the country.

6.3.5 Disability and Metaverse

The second most seen issue in diversity debates concerns users with disabilities. Erken (2022) was already quoted previously. Virtual reality and augmented reality programming are primarily visual mediums (Analytics Insights, 2022), and visually disabled citizens are left behind in the initial stages of metaverse. 3D-audio echo-location, audio descriptive menus, and haptic feedback are listed as some technical remedies. The *Analytics Insights* piece describes other types of disabilities too. Sensory enhancement tools may help citizens with hearing or visual weaknesses. Project VOISS[8] (Virtual Reality Opportunities to Integrate Social Skills) utilises virtual reality to re-enact social communication and assist individuals with ASD (autism spectrum disorder). In another case, a project from Duke University, Walk Again Project,[9] uses brain-machine interfaces and a virtual reality system to help people with paraplegia. Thus, there are some early attempts to make metaverse more accessible to users with disabilities.

6.3.6 Appeal to Corporations

Most writing on diversity may be classified under the appeal to corporations. As noted previously, leadership in metaverse platforms lacks diversity. The benefits of diversity are highlighted, and the lack of diversity is presented as one of the factors that may lead a company to fail in its business. When management does not see diversity and inclusion as a priority, a targeted group, such as people of colour, women, femmes, disabled people, religious minorities, lesbians, gays, bisexuals, and transgender people, will be marginalised in the workplace (McKinsey & Company, 2020). Engagement with diversity should be authentic, not another kind of -washing (like greenwashing, brainwashing, etc.) (Georgiou, 2021).

Eddy (2022) highlights that while companies invest in their products, they leave some team members by not investing in the infrastructure to be inclusive. Companies should work on a safe environment for VR and AR usage for the users and employees.

A tactic to persuade businesses is to focus on the benefits of diversity. A metaverse that is meaningful for everyone can happen only with creators from diverse backgrounds. It is emphasised that not the technology but the new people brought into the metaverse will be more critical (Fast Company, 2022). Sumagaysay (2022) argues that companies with more non-white workers in management roles produce more substantial cash flow, profit, revenue, return on equity, and stock performance. His arguments are based on an analysis of data from nearly 300 companies. The study claims that there is a positive link between diverse workforces and management and the financial performance of companies.

Similarly, Putriani and Aras quote that ethnic/cultural diversity in executive teams is 33 per cent more likely to generate industry-leading profitability (2022). A more inclusive and diverse workforce makes an organisation more productive, attracts the right talent, and creates societal benefits; gender-inclusive teams were better at CSR efforts and contributed more to charitable funds (Sachdeva, 2022).

The stats about diversity rations may be negative, but many corporations seem responsive to these calls. Meta explicitly announced to facilitate more inclusion and opportunity by recruiting a more diverse set of creators and experts (Hutchinson, 2022). It strongly emphasises diversity in the creator base that will help to include historically excluded sectors in the digital sphere. The company has a special chief diversity officer, and it aims to prove its sincerity through a series of practices. To recruit a more diverse set of talents, the company goes beyond the traditional hiring methods. It has partnered with Women in Immersive Tech Europe[10] to fight gender disparity and released funds for academic research on women's safety issues. Women are the builders of many filters in Meta's Spark AR production studios (Pantony, 2022). In the meantime, the company allows creating an avatar in more than one quintillion different ways. The company is working on providing more options, such as wheelchairs, while ear implants and tools are built to translate languages so that non-English speaking users can benefit (Bacchus, 2022). Meta also announced to work with global organisations such as the Organisation of American States (OAS) to build the metaverse responsibly. It started to hire PhDs in machine learning and artificial intelligence in Latin America for the first time in 2022 (Bullfrag Blog, 2022). Its inclusive policies are evidenced by groups such as the Gatherverse[11] and the Virtual and Augmented Reality Association (VRARA)[12] (XR Today, 2022).

Both Meta and Lenovo began to share their gender baselines. The metrics are welcome as benchmarks for other companies but also provide data for diversity calculations (Lenovo, 2022). Other companies in the field are also experimenting or collaborating with civil society. CRADL[13] partnered with organisations – like Black Bitcoin Billionaire, HER DAO, the Africa Blockchain Institute, and IMPAQTO – to involve minority communities in Web3 projects (Kuhn, 2022). The Institute of Digital Fashion teamed up with Daz 3D to release the world's first non-binary digital double (Bacchus, 2022). Yahye Siyad of Cyber-Duck[14] explains the difficulties in reaching diversity in tech industries but higher success rates of diversity norms' acceptance among new employees (Blackwell, 2022).

6.3.7 Risks and Remedies

Metaverse can become a potent tool of persuasion without proper regulations (Rosenberg, 2022). An unprecedented collection of new data types (including biometric and visual data) and algorithmic hyper-engagement (Duan et al., 2021) may lead to mass persuasion and the reproduction of existing inequalities. The rapid growth of metaverse may reduce multiculturalism and diversity, and the already occurring sexual assault and harassment of female avatars will increase (Ambolis, 2022). The potential for real-life harassment and abuse spilling into the metaverse is reported, and Europol has already warned about the negative impacts of these cases' increasingly realistic XR experiences (Carter, 2022). Another research based on a social media scan of thousands of conservations is that consumers are concerned with sexual harassment, personal safety, and privacy issues (Capgemini, 2022). A few early incidents also fed these concerns. Among them, in one of the highly circulated ones, a user in Meta's Horizon Venues claimed to be gang-raped (Frishberg, 2022). A researcher claimed to be harassed too (Soon, 2022).

After the incidents mentioned previously, Meta made technical changes to prevent similar incidents. In the meantime, content moderation emerged again as a critical issue. Even with AI usage, social media platforms have difficulties handling content moderation (Gillespie, 2020), and metaverse appears as a possibly more complex place. The trolling in metaverse is more visceral than in social media regarding what one feels, and moderation is more complicated (Mak, 2022). Hanna (2022) argues that the more immersive the virtual world, the more potential for invasive and damaging abuse. In this context, Duan et al. (2021) propose a three-layer architecture for metaverse that includes infrastructure, interaction, and ecosystem. This explicit model intends to prevent technological solutionism and include social aspects in the platform building.

Metaverse platforms are constantly urged to prepare for a safe environment so that diversity can be welcome. However, social media experience demonstrated that regulations must be beyond the platforms' self-control mechanisms. For instance, the EU's General Data Protection Regulation (GDPR) should be revised to deal with the metaverse. Thus users will be adequately protected and informed against the previously unimagined types of real-time data collection (Anwar, 2022).

A final call may be related to the educational field. Garivaldis et al. (2022) mention the importance of diversity in learning. Their edited volume is on digital learning in general, and metaverse does not occupy a significant part, but diversity is strongly implied in learning environments to train future generations. Wang and Medvegy (2022) are more vocal in the role of teachers: Teachers will be social facilitators and program and train AI. Technology skills should be taught along with soft skills such as empathy, creativity, communication, and collaboration.

Risks and remedies here are limited to their connection to diversity issues, and there will probably be a growing literature on this shortly.

6.4 CONCLUSION

Early research points out the importance of defining principles and practices in building the metaverse. It is critical to start with safety, inclusiveness, and accessibility to guarantee diversity in metaverse (Zallio & Clarkson, 2022). I agree with Zallio and Clarkson that new skills and access technologies are needed to ensure a diversity of users in metaverse. This starts in the very building process. The current ratios may be upsetting, but the concern over diversity is a frequent theme, and many corporations take diversity policies seriously.

It is also vital to develop diversity programs productively. Putriani and Aras (2022) argue that these programs fail if organisational specifics, emotional attitudes, tailored approaches to managerial positions, and many other variables are ignored.

In conclusion, if the metaverse building occurs with a human-centric orientation, seemingly utopian expectations can be realised:

> Since users in the Metaverse will interact in the digital space as virtual images, many problems in the real world can be avoided. The advantages of Metaverse applied to smart cities include (1) better accessibility so that users in different physical locations can enjoy the same information and experience; (2) better diversity so that different user groups can enjoy a space to get along with each other free from physical resources; (3) better equality so that users of different races, colours, and languages can enjoy equal opportunities for development; (4) better humanity so that human culture to be passed on more healthily and perpetually. More importantly, the Metaverse may be an essential infrastructure for future technological innovation.
>
> (Wang & Medvegy, 2022)

NOTES

1 This definition is based on National Center for Education Statistics website (https://nces.ed.gov/).
2 Diversity, Equity and Inclusion (DEI) is a frequently used term in the corporate speak (Arsel et al., 2022).
3 In the UK, 16% of cyber security professionals are female (Leonard, 2022)
4 *Hypebae*. (2022, September 20). QUALIA launches Web3 incubators new creatives. Retrieved January 2, 2023, from https://hypebae.com/2022/9/qualia-web3-incubator-creatives-metaverse
5 BFF: Empowering friends who crypto. (n.d.). Retrieved January 2, 2023, from www.mybff.com/
6 World of Women: Home. (n.d.). Retrieved January 2, 2023, from www.worldofwomen.art/
7 Web3 Women Union|looking for core contributors – Twitter. (n.d.). Retrieved January 2, 2023, from https://mobile.twitter.com/web3womenunion
8 Project VOISS – Giving Students a Voice. (n.d.). Retrieved January 3, 2023, from www.projectvoiss.org/
9 The Walk Again Project. (n.d.). Retrieved January 3, 2023, from www.walkagainproject.org/
10 Women in Immersive Tech. (n.d.). Retrieved January 3, 2023, from www.wiiteurope.org/
11 GatherVerse – A Step Closer to the Metaverse. (n.d.). Retrieved January 4, 2023, from https://gatherverse.org/
12 VR/AR Association – The VRARA. (n.d.). Retrieved January 4, 2023, from www.thevrara.com/

13 CRADL – Crypto Research and Design Lab Web3 Education: Home. (n.d.). Retrieved January 4, 2023, from https://cradl.on360.co/
14 Cyber-Duck: Award Winning Digital Agency. (n.d.). Retrieved January 4, 2023, from www.cyber-duck.co.uk/

REFERENCES

Alaghband, M., & Yee, L. (2022, October 21). Even in the Metaverse, women remain locked out of leadership roles. *McKinsey.* www.mckinsey.com/featured-insights/diversity-and-inclusion/even-in-the-metaverse-women-remain-locked-out-of-leadership-roles

Ambolis, D. (2022, December 9). How will the metaverse affect the future of work? *Blockchain Magazine.* https://blockchainmagazine.net/how-will-the-metaverse-affect-the-future-of-work/

Analytics Insights. (2022, March 11). Metaverse is the future: Will it help disabled people live better? *Analytics Insight.* www.analyticsinsight.net/metaverse-is-the-future-will-it-help-disabled-people-live-better/

Anwar, M. J. (2022, October 20). *How will digital identity shape the Metaverse?* https://fintechmagazine.com/articles/how-will-digital-identity-shape-the-metaverse

Arsel, Z., Crockett, D., & Scott, M. L. (2022). Diversity, equity, and inclusion (DEI) in the Journal of Consumer Research: A Curation and Research Agenda. *Journal of Consumer Research, 48*(5), 920–933.

Averstad, P., Baboolall, D., Beltrán, A., Lefkowitz, E., Nee, A., Pinshaw, G., & Quigley, D. (2022). *The state of diversity in global private markets.* McKinsey & Company.

Bacchus, M. (2022, October 12). *The ethical implications of work in a digital world.* www.hrgrapevine.com/content/article/2022-11-18-diverse-in-the-metaverse-ethical-implications-of-a-digital-world

Barnes, T. J., & Wilson, M. W. (2014). Big data, social physics, and spatial analysis: The early years. *Big Data & Society, 1*(1).

Beer, D. (2009). Power through the algorithm? Participatory web cultures and the technological unconscious. *New Media & Society, 11*(6), 985–1002.

Blackwell, L. (2022). If you can't change it, grow around it: The leaders rallying for diversity and inclusion. *The Drum.* Retrieved December 22, 2022, from www.thedrum.com/news/2022/10/31/if-you-can-t-change-it-grow-around-it-the-leaders-rallying-diversity-and-inclusion

Brandusescu, A., & Reia, J. (2022). *Artificial intelligence in the city: Building civic engagement and public trust.* Centre for Interdisciplinary Research on Montreal, McGill University.

Bullfrag Blog. (2022, October 20). *Metaverse has the potential to promote diversity and inclusion.* www.bullfrag.com/metaverse-has-the-potential-to-promote-diversity-and-inclusion/

Capgemini. (2022). *Nine in ten consumers are curious about the Metaverse.* https://outsourcecommunications.prezly.com/nine-in-ten-consumers-are-curious-about-the-metaverse

Carah, N., & Angus, D. (2018). Algorithmic brand culture: Participatory labour, machine learning and branding on social media. *Media, Culture & Society, 40*(2), 178–194.

Carter, R. (2022, November 16). Will the Metaverse need policing? *XR Today.* www.xrtoday.com/virtual-reality/will-the-metaverse-need-policing/

Consultancy.uk. (2022, November 28). *Women shut out of leadership roles in the Metaverse.* www.consultancy.uk/news/32903/women-shut-out-of-leadership-roles-in-the-metaverse

Dalton, C. M. (2013). Sovereigns, spooks, and hackers: An early history of Google geo services and map mashups. *Cartographica: The International Journal for Geographic Information and Geovisualization, 48*(4), 261–274.

Datta, A. (2022, October 27). In its simplest form, metaverse is the digital extension of a smart city. *Geospatial World.* www.geospatialworld.net/prime/in-its-simplest-form-metaverse-is-the-digital-extension-of-a-smart-city/

Digital Trends. (2022, October 20). Big Tech's metaverse vision is weak. Here's what it needs. *Digital Trends.* www.digitaltrends.com/computing/what-the-metaverse-is-missing/

Dobbin, F., & Kalev, A. (2013). The origins and effects of corporate diversity programs. In Q. M. Roberson (Ed.), *The Oxford handbook of diversity and work* (pp. 253–281). Oxford University Press.

Dobransky, K., & Hargittai, E. (2016). Unrealised potential: Exploring the digital disability divide. *Poetics, 58,* 18–28.

Duan, H., Li, J., Fan, S., Lin, Z., Wu, X., & Cai, W. (2021). Metaverse for social good: A university campus prototype. In *Proceedings of the 29th ACM international conference on multimedia* (pp. 153–161). Association for Computing Machinery.

Dupuy, E. (2022, October 20). Council post: Learning from the past to build a safe and inclusive Metaverse for the future. *Forbes.* www.forbes.com/sites/forbesbusinessdevelopmentcouncil/2022/10/20/learning-from-the-past-to-build-a-safe-and-inclusive-metaverse-for-the-future/

Eddy, N. (2022, November 1). Into the Metaverse: Making the case for a virtual workspace. *InformationWeek.* www.informationweek.com/big-data/into-the-metaverse-making-the-case-for-a-virtual-workspace

Eisenstat, Y. (2019, February 12). The real reason tech struggles with algorithmic bias. *Wired.* www.wired.com/story/the-real-reason-tech-struggles-with-algorithmic-bias/

Ely, R. J., & Thomas, D. A. (2020). *Getting serious about diversity: Enough already with the business case.* https://hbr.org/2020/11/getting-serious-about-diversity-enough-already-with-the-business-case

Engler, A. (2022, June 8). The EU AI Act will have global impact, but a limited Brussels Effect. *Brookings.* www.brookings.edu/research/the-eu-ai-act-will-have-global-impact-but-a-limited-brussels-effect/

Erken, F. (2022). Sosyo-dijital eşitsizlik ekosisteminde sanal gerçeklikten metaverse'e erişilebilirlik: içeridekiler ve dişaridakiler. *Ufkun Ötesi Bilim Dergisi, 22*(1), 84–99.

Estrada, S. (2022, October 22). Even the Metaverse has a lack of women in leadership roles. *Fortune.* https://fortune.com/2022/11/22/even-the-meta verse-has-a-lack-of-women-in-leadership-roles/

Fast Company. (2022, October 20). *How creators are making an inclusive metaverse.* www.fastcompany.com/90798140/how-creators-are-making-an-inclusive-metaverse

Frishberg, H. (2022, January 1). Mom opens up about being 'virtually gang raped' in Metaverse. *NY Post.* https://nypost.com/2022/02/01/mom-opens-up-about-being-virtually-gang-raped-in-metaverse/

Garivaldis, F., McKenzie, S., Henriksen, D., & Studente, S. (2022). Achieving lasting education in the new digital learning world. *Australasian Journal of Educational Technology, 38*(4), 1–11.

Georgiou, M. (2021, March 15). Council post: How and why to build brand authenticity. *Forbes.* www.forbes.com/sites/forbescommunicationscouncil/2021/03/15/how-and-why-to-build-brand-authenticity/

Gillespie, T. (2020). Content moderation, AI, and the question of scale. *Big Data & Society, 7*(2).

Haklay, M. (2013). Neogeography and the delusion of democratisation. *Environment and Planning A, 45*(1), 55–69.

Hanna, S. (2022, November 16). Why content moderation could make or break the Metaverse. *Fast Company.* www.fastcompany.com/90811476/why-content-moderation-could-make-or-break-the-metaverse

Hollowell, A. (2022, December 8). What layoffs mean for a tech industry that already lacks diversity. *VentureBeat.* https://venturebeat.com/ai/what-layoffs-mean-for-a-tech-industry-that-already-lacks-diversity/

Horwood, P. (2022, November 10). *Discussing the possibilities of a diverse metaverse.* www.computing.co.uk/news/4059977/discussing-possibilities-diverse-metaverse

Hutchinson, A. (2022, October 19). Meta paints a picture of an idealistic metaverse in new future planning report. *Social Media Today.* www.socialmediatoday.com/news/Meta-shares-insights-into-metaverse-future-report/634519/

Infosys. (2022). *Infosys receives top honors for diversity, equity and inclusion practices.* www.infosys.com/newsroom/features/2022/top-honors-diversity-equity-inclusion-practices.html

Kitchin, R. (2014). Big Data, new epistemologies and paradigm shifts. *Big Data & Society, 1*(1), 2053951714528481.

Kitchin, R., & Dodge, M. (2011). *Code/space: Software and everyday life.* MIT Press.

Kuhn, D. (2022, October 26). *What a Web3 hackathon teaches us about diversity in crypto.* www.coindesk.com/layer2/2022/10/26/what-a-web3-hackathon-teaches-us-about-diversity-in-crypto/

Lacity, M., Mullins, J. K., & Kuai, L. (2022). *What type of Metaverse will we create?* University of Arkansas.

Lenovo. (2022, October 21). USIPA applauds diversity pledge update. *AccessWire.* www.accesswire.com/721581/

Leonard, J. (2022, October 21). *UK cyber workforce grows 12% but there's still a 73% shortfall, report.* www.computing.co.uk/news/4058608/uk-cyber-workforce-grows-theres-shortfall-report

Mager, A. (2018). Internet governance as joint effort: (Re)ordering search engines at the intersection of global and local cultures. *New Media & Society, 20*(10), 3657–3677. https://doi.org/10.1177/1461444818757204

Mak, A. (2022, May 9). I was a bouncer in the Metaverse. *Slate.* https://slate.com/technology/2022/05/metaverse-content-moderation-virtual-reality-bouncers.html

McClain, J. (2022, November 5). Expect real-world beauty standards to stay the same in the Metaverse. *Rolling Out.* https://rollingout.com/2022/11/05/expect-real-world-beauty-standards-to-stay-the-same-in-the-metaverse/

McKinsey & Company. (2020, June 23). How organisations can foster an inclusive workplace. *McKinsey.* www.mckinsey.com/capabilities/people-and-organizational-performance/our-insights/understanding-organizational-barriers-to-a-more-inclusive-workplace

Pantony, A. (2022, March 8). 'Where do women fit into the metaverse?': This is what went down during GLAMOUR's first meeting in the Metaverse. *Glamour UK.* www.glamourmagazine.co.uk/article/glamour-meeting-metaverse

Parihar, N. (2022, November 21). Can AI eliminate bias and promote diversity and inclusion? *People Matters.* www.peoplematters.in/article/diversity/can-ai-eliminate-bias-and-promote-diversity-and-inclusion-36101

Peyton, L. (2022, October 20). Manifesto for the makers of the Metaverse. *MarTech.* https://martech.org/manifesto-for-the-makers-of-the-metaverse/

Press Release Network. (2022, December 4). Press Release Network – a cyber gear initiative. *Press Release Network.* https://pressreleasenetwork.com/site/2022/12/14/gen-z-and-the-metaverse/

Putriani, I., & Aras, M. (2022). Elevate company's social media as a diversity, equity, and inclusion platform: Multi cases study to digital technology companies. *Journal of World Science, 1*(10), 906–916.

Ramirez, E. G. (2021). Diversity, equity, and inclusion: Is it just another catch-phrase? *Advanced Emergency Nursing Journal, 43*(2), 87–88.

Responsible Metaverse Alliance. (2022). Charter. *Responsible Metaverse Alliance.* https://responsiblemetaverse.org/charter/

Rosenberg, L. (2022, October 22). Mind control: The Metaverse may be the ultimate tool of persuasion. *VentureBeat.* https://venturebeat.com/virtual/mind-control-the-metaverse-may-be-the-ultimate-tool-of-persuasion/

Ruelas, B. (2022, November 7). Dr. Tiffany Brandreth discusses a new future for diversity, equity, and inclusion. *Grit Daily News.* https://gritdaily.com/dr-tiffany-brandreth-discusses-a-new-future-for-diversity-equity-and-inclusion/

Ryder, B. (2022, November 6). Move over crypto bros, it's time to meet the women of web3. *Jing Daily.* http://jingdaily.com/crypto-bros-women-of-web3/

Sachdeva, J. (2022, December 13). Why data and tech are crucial to workplace diversity. *People Matters.* www.peoplematters.in/article/diversity/why-data-and-tech-are-crucial-to-workplace-diversity-36326

Saka, E. (2020). Big data and gender-biased algorithms. In *The international encyclopedia of gender, media, and communication* (p. na). Wiley Blackwell.

Schijns, J. (2022, December 6). Why the Metaverse is an opportunity for women in tech. *Acceleration Economy*. https://accelerationeconomy.com/digital-business/why-the-metaverse-is-an-opportunity-for-women-in-tech/

Shardeum. (2022, September 19). 6 Types of decentralisation – You should know. *Shardeum. EVM Compatible Sharded Blockchain*. https://shardeum.org/blog/what-are-the-types-of-decentralization/

Sharma, M. (2022, November 17). "Diversity deficit": Tech talent market still male-dominated. *People Matters*. www.peoplematters.in/article/diversity/diversity-deficit-tech-talent-market-still-male-dominated-36064

Shorey, S., & Howard, P. (2016). Automation, big data and politics: A research review. *International Journal of Communication, 10*.

Soon, W. (2022, May 30). A researcher's avatar was sexually assaulted on a metaverse platform owned by Meta, making her the latest victim of sexual abuse on Meta's platforms, watchdog says. *Business Insider*. www.businessinsider.com/researcher-claims-her-avatar-was-raped-on-metas-metaverse-platform-2022-5

Stebbins, R. A. (2001). *Exploratory research in the social sciences* (Vol. 48). Sage.

Sumagaysay, L. (2022, November 17). Racial diversity in management linked to positive financial performance, new analysis shows. *MarketWatch*. https://www.marketwatch.com/story/racial-diversity-in-management-linked-to-positive-financial-performance-new-analysis-shows-11668693795

Sun, J., Gan, W., Chen, Z., Li, J., & Yu, P. S. (2022). *Big data meets Metaverse: A survey*. Cornell University. https://arxiv.org/abs/2210.16282

Thatcher, J. (2014). Big data, big questions| Living on fumes: Digital footprints, data fumes, and the limitations of spatial big data. *International Journal of Communication, 8*, 1765–1783.

Van Dijck, J. (2013). *The culture of connectivity: A critical history of social media*. Oxford University Press.

Wang, J., & Medvegy, G. (2022, October 13–17). Exploration the future of the Metaverse and smart cities. In *Proceedings of the 22th international conference on electronic business*. ICEB.

XR Today. (2022, December 14). Meta's responsible Metaverse and the enterprise. *XR Today*. www.xrtoday.com/virtual-reality/responsible-metaverse-meta-platforms/

Zallio, M., & Clarkson, P. J. (2022). Designing the Metaverse: A study on inclusion, diversity, equity, accessibility and safety for digital immersive environments. *Telematics and Informatics, 75*, 101909.

Zarrin, J., Wen Phang, H., Babu Saheer, L., & Zarrin, B. (2021). Blockchain for decentralisation of internet: Prospects, trends, and challenges. *Cluster Computing, 24*(4), 2841–2866.

Zhao, Y., Bennett, C. L., Benko, H., Cutrell, E., Holz, C., Morris, M. R., & Sinclair, M. (2018). Enabling people with visual impairments to navigate virtual reality with a haptic and auditory cane simulation. In *Proceedings of the 2018 CHI conference on human factors in computing systems* (pp. 1–14). Association for Computing Machinery.

CHAPTER 7

Queering the Metaverse

Queer Approaches to Virtual Reality in Contemporary Art

Vítor Blanco-Fernández, MA. Predoctoral Fellow

Universitat Pompeu Fabra, Spain

CONTENTS

7.1 INTRODUCTION

In August 2022, Mark Zuckerberg launched *Horizon Worlds*, Meta's most renowned VR social network, in France and Spain simultaneously (Mehta, 2022). To announce it, Zuckerberg posted a picture of a 3D rendering of both Paris' Tour Eiffel and Barcelona's Sagrada Familia. The image quickly became an Internet phenomenon due to its low quality, coming to prove that, besides billionaire investments in the metaverse, Zuckerberg's envision of the virtual realm is one of poor aesthetics and empty spaces.

DOI: 10.1201/9781003379119-7

Furthermore, this is not the only controversy surrounding *Horizon Worlds*, from laugh-provoking announcements such as "Meta figured out legs for its avatars" (Peters, 2022), to denounces of virtual sexual harassment (Morla, 2022).

According to Andrew Durbin (2022), this demonstrates "how impoverished Silicon Valley's progress has been" (p. 13). On the contrary, "the most exciting innovations in digital landscapes, technological futures and social spaces can be found in artists' studios rather than developers' dens in Menlo Park" (p. 13). Durbin is not alone in this idea. Anna Munster (2006) underlines contemporary artists "using VR, the Internet and mixed-reality media . . . to create a form of drifting that not only signals a different kind of movement through, but also a different production of, digital spaces" (p. 103). And Sumugan Sivanesen (2017) suggests that contemporary artists are the most important workforce against Meta's monopoly of the metaverse.

This is particularly close to queer theorist José Esteban Muñoz's description of how art helps us "to see the not-yet-conscious" (2009, p. 3). According to Muñoz, art is the arena of world-making. In art, we can find alternatives to the quotidian and the normative, especially for queer, BIPOC people. Applying Muñoz's queer world-making (2009) to the metaverse (Blanco-Fernández, 2022), implies searching for digital alternatives to Meta's monopoly, in contemporary art, particularly through the work by queer, BIPOC artists. Altogether, this chapter questions how queer artists are imagining the metaverse otherwise. By conducting a close reading analysis of five artworks, it aims to demonstrate that through their queer world-making, we can glimpse what a different virtuality can look like.

7.2 THE "INDUSTRIAL CONTINUUM" OF THE METAVERSE

In *Volumetric Regimes*, Jara Rocha and Femke Snelting (2022) named the "industrial continuum of 3D" to the interchangement of three-dimensional modelling developments through disciplines "such as bio-medical imaging, wildlife conservation, border patrolling, and Hollywood computer graphics" (Rocha & Snelting, 2022, p. 219). This includes, of course, metaverse development. One of its examples is the industrial continuum of video game engines, as described by Luca Carrubba in the upcoming book *Videogame Trouble* (2023). Carrubba accounts how two engines – Unity3d and Unreal Engine – add up to 56% of the market share. Simultaneously, their tools "go beyond the creation of games to be increasingly used in areas such as architecture, urban planning, and education, among others" (Carrubba, 2023), monopolistically deciding how volumes look in all these disciplines. This is what Rocha and Snelting (2022) define as "volumetric regimes", normative ways of rendering that leave "very little space for radical experiments and surprise combinations" (Rocha & Snelting, 2022, p. 219).

Additionally, these volumetric regimes become particularly perverse when noticing its connections to resource extraction, colonial heritage, and patriarchy. Regarding colonialism, Internet infrastructures, unequal distribution of access to technology, and tech-garbage dumps, mirror former relations of Western imperialism (Rezaire, 2022). As for capital, venture capital institutions decide how technology looks like from a neoliberal ideology (Blas, 2022), with a particular interest in monitoring and selling our data – also studied from a gender perspective (D'Ignazio & Klein, 2020). Finally, gender lenses to virtuality have pointed out misogynist and racist depictions of avatars (Bennett & Beckwith, 2018), as well as queer antagonist, misogynist, or racist attitudes in the metaverse (Leonard, 2020; Morla, 2022).

The industrial continuum of the metaverse is easy to localise. First, by renaming its enterprise to Meta, Zuckerberg succeeded in connecting its business branding to the future of virtuality. Second, by buying Oculus, one of the most important designers and sellers of user-friendly virtual reality head-mounted displays, Meta controls both the possibility of creating mainstream metaverses, as well as their doors. This becomes especially triggering when considering Facebook's record of breaking privacy rights (Isaak & Hanna, 2018), misinformation campaigns (Vaidhyanathan, 2021), and the Cambridge Analytica scandal (Hinds et al., 2020).

Meta's monopoly of the metaverse is only contested by the AAA video game industry. This translates into a specific way of building the metaverse, from aesthetical decisions to usability, user experience or broader ways of understanding virtuality and space. Regarding aesthetics, we find two paradigmatic lines: 1) hyperrealist modelling of three-dimensional environments and figures, especially in AAA videogames (Lanier, 2022); and 2) cartoonish aesthetics in "cute" environments and simplified avatars (Stokel-Walker, 2022), such as the ones in *Horizon Worlds*, as well as *VRChat*. Regarding the depiction of the space, we find how cartesian laws of perspective are maintained (Stone, 1991), inheriting Western pictorial tradition (Tavinor, 2022). Hito Steyerl (2011) denounces the cartesian depiction of space for its epistemological connection with colonialism. The humanist Cartesian horizon, the author notes, is connected to slavery ships: a single point of view, a linear time, a "calculable, navigable, and predictable" space and future. All of them worked as "an additional toolkit for enabling Western dominance, and the dominance of its concepts" (Steyerl, 2011). Mirroring Cartesianism in virtual scenarios, Steyerl suggests, implies unwittingly mirroring these problematic inheritances. Related to the space and the horizon is Sara Ahmed's critique of orientation (2006). Ahmed considers mainstream orientations as imposed life paths which are violent for plenty of people, especially women, queer folk, and/or BIPOC. In virtual worlds, especially video games, developers design predefined paths, suggested or mandatory (Calleja, 2011), thus codifying orientation. On the contrary, Rocha and Snelting (2022, pp. 57–74) search for "disorientating" volumetric representations that help us "relate to a world that is becoming oblique, where inside and outside, up and down switch places and where new perspectives become available" (p. 58).

Another mainstream metaverse trait is its obsession with reality. Although certain fantasy elements are welcomed in these spaces, simultaneously, investments are prone to duplicate the "outside" world. *Horizon Worlds'* Tour Eiffel and Sagrada Familia are two examples, as well as Zuckerberg's Instagram post from June the 10th, 2022, where he described Meta's goal as making "the Metaverse as realistic as the physical world". Particularly connected to reality obsession are avatars and digital identities. On the first hand, researchers from the Queer Game Studies field (Ruberg & Shaw, 2017) have described how avatar making, although increasingly more inclusive, still depends on a hermetic idea of identity as fixed checkboxes, which Galloway defines as "modular, instrumental, typed, numerical, algorithmic" (2006, p. 102). On the other hand, and considering digital apps developers' obsession with authenticity – Facebook's or Tinder's identity checks are good examples – we can also predict that metaverse's avatars will be defined by a conservative idea of "true" identity shortly (Lanier, 2022, p. 96).

The idea of "truthiness" comes connected to metaverse's commodification of social norms and conservative values, censoring, hiding, or banning radical self-expression. We can observe this trend toward normalcy if we compare *Horizon Worlds* or *VRChat* to previous metaverses such as *Second Life* and its spectrum of queer users and practices, including virtual sex (cárdenas, 2010). Also in body representation, especially in the study of how three-dimensional modelling programs, such as *MakeHuman* and *Meta Human*, decide what we can and cannot do technologically when designing a human body, consequently defining humanness (Rocha &

Snelting, 2022, pp. 191–198). Finally, we should acknowledge that the mainstream metaverse is a business-oriented model. This implies sponsored content, data extraction, NFTs, crypto, or traditional currency business, as well as payment walls that separate users spatially depending on their economic capacity.

7.3 THINKING THE METAVERSE (QUEERLY) OTHERWISE

To suggest alternatives to the mainstream metaverse, here I analyse five works of art done by queer artists. The sample decision is purely purposeful (Suri, 2011) and responds to two requisites: 1) they are connected to the metaverse aesthetics (three-dimensional forms, world-making...) and technologies (avatar embodiment, use of virtual and extended realities...), and 2) they are done by queer artists, with particular emphasis on trans and BIPOC artists. Consequently, the sample looks as follows:

- *Ressurrection Lands*, Danielle Brathwaite-Shirley (2021)

- *Domestika*, Jacolby Satterwhite (2017)

- *Sin Sol/No Sun*, micha cárdenas (2018)

- *Vampyroteuthis Infernalis*, Pete Jiadong Qiang (2020)

- *Sick Trans-sex Gloria*, Tabitha Nikolai (2017)

The analysis is based on the close reading method (Brummett, 2018). Close reading analysis emerged in literary studies but has been applied widely elsewhere, including interactive, virtual discourses, such as video games (Bizzocchi & Tanenbaum, 2011). It consists in a deep, critical, and interconnected analysis of cultural products, not only considering their content but their broader impact and their position in the social and epistemological context they emerge. Close reading allows us to focus on specific elements, purposefully signalling those that better illustrate the problem we are questioning. Here, my close reading focuses on five main traits that help us imagine the metaverse otherwise: aesthetics, storytelling, world-making, avatar/identity, and interactivity.

7.3.1 Aesthetics

Hyper-realism in computer-generated images, defined as the detailed, perfectly lightened, physically accurate modelling, is absent from the works analysed. Cartoonish aesthetics are also missing. Nikolai's *Sick Trans-sex*

Gloria is the most realistic depiction, but its 2000's graphics – resembling *The Sims 2* or *Second Life* – make it retro. Meanwhile, cárdena's *Sin Sol/No Sun* is an extended reality, thus part of its aesthetic experience comes from the place we are standing when visiting it. Although the digitally added characters are also 2000-ish, cárdenas introduces other aesthetic traits, particularly glitch effects. Following *Glitch Feminism* (Russell, 2020), queer failure (Halberstam, 2011), and her proposal of queer prototyping (cárdenas, 2010), her glitch aesthetics suggest that there are fruitful possibilities in mistakes. Aura, a trans-Latina AI, appears in our phone "giving the impression that a glitch might have caused her arrival" (Casalini, 2021, p. 8). And through this unplanned encounter, we understand the consequences of climate change on migrant, trans, and BIPOC people.

The other three pieces are also examples of rendering aesthetics thematically, that is, voluntarily modifying realism as a narrative resource. Jacolby Satterwhite and Pete Jiadong Qiang invest in maximalism. Satterwhite is particularly renowned for his maximalist "science-fictional glitch worlds" (Harris, 2021, p. 6). *Domestika* is a mix of modelling and recorded elements (dancing queer bodies), renderings of his mother's drawings (who suffered from dementia), neon signs, queer-coded elements including "human-like figures in chains linked to white male bodies dressed in sadomasochistic gear, flying Pegasi ridden by dark figures, topless femme cyborgs"; all "inserted into a surreal, futuristic landscape that can be best described as part industrial nightclub, part space station, and part domestic space" (Misra, 2020, p. 41). Qiang does not reach Satterwhite's chaos. His composition is based on what he names ACGN – meaning anime, comic, game and novel – all together with internet elements and fan communities (Figure 7.1). Another important element in Qiang's work is thematic texturing: using non-realistic textures to transmit a particular message or aesthetic experience. In *Vampyroteuthis Infernalis*, thematic texturing implies "3D scans of spaces, cartographies, and surfaces in London and Berlin", which "are remixed into multiple worlds within the game engine" (Transmediale, 2022). Nevertheless, Danielle Brathwaite-Shirley is the main exponent of queer thematic texturing. *Resurrection Lands* draws from first-generation TV consoles as a political response to hyperrealism (Khanna & Brathwaite-Shirley, 2021), and its textures come from Black and trans communities: "skies are made from skin, grass from hair, clothes from old family photographs" (Thomas, 2022).

Finally, it is important to notice that the five pieces include text as a graphic element. While social and gaming virtual worlds try to substitute

FIGURE 7.1   Screenshot of *Vampyroteuthis Infernalis* by Pete Jiadong Qiang.
Source: Courtesy of the artist.

text-based information for other immersive elements (such as audio or narrative development), the artists from this sample trust in text-based graphics as an important element of their metaverse proposals.

7.3.2 Storytelling

While mainstream metaverses are inclined towards social networking, within this sample we find complex storytelling around contemporary topics, such as racism and colonialism, transphobia, queer communities, family, climate change, apocalypse, body, or identity. Plots range from abstract to classic storytelling. On the abstract side, Satterwhite's *Domestika* has been defined as "non-narrative" (Misra, 2020), although family, home, young queerness, sex, or science fiction can be interpreted as concrete themes. Qiang's *Vampyroteuthis Infernalis* is abstract as well, although it portrays the "HyperBody", a concept developed by the artist to describe the entanglement of bodies and digital contemporary cultures.

Brathwaite-Shirley's *Resurrection Lands* follows an interactive storytelling device to reflect on Black trans memory, archive, and ancestry. After building a Black and trans digital archive, white cis people conquered it and converted it into an e-sports game, playing the "Black trauma" (Hart, 2020). Consequently, the Black trans ancestors decided to put an access policy: the experience you have in the *Resurrection Lands* will change considerably depending on if you are Black and trans, or white and cis.

From respect, care, and love, to lack of empathy, reparation, or account-ability. Through its six stages – Hormone Bar, Whispers, Trans Exhibition, Ancestors Worlds, The Lake, and Deadname Burial – topics such as ances-try, community, care, memory, ocean (and slavery), death, deadnames, embodiments, gaze politics, hormones, reparation, gender binarism, or transnormativity, emerge.

The last two pieces address apocalypse and science fiction. cárdenas' *Sin Sol/No Sun* is set 50 years from now, when Aura, a trans-Latina AI, informs us of the massive wildfires caused by climate change that ended her world. Moreover, it is a story about xenophobia and migrant racist policies: the migrants escaping from climate change found police brutality and border surveillance when trying to cross the frontiers. Aura's only way of escaping the wildfires was by coding herself into the digital realm. Therefore, main-stream transhumanism is substituted for Afrofuturism or Afro-fabulation (Nyong'o, 2019): the Black trans-Latina survives (digitally), rather than the white, cisgender, male, and tech-entrepreneur. Nikolai's *Sick Trans-sex Gloria* is also a story of queer survival (Kouri-Towe, 2014) and trans-futurism (Asante et al., 2021). Gloria survives the apocalypse by inhabiting a solitary island covered by a soy plantation. By cultivating soy, Gloria gets the hormones she needs, connecting this tale with traditions of DIY hor-mones and biohacking within the trans community.

7.3.3 World-Making

The worlds depicted in the analysed works – except *Sin Sol/No Sun*, as an XR piece – are not mirrors of reality, and they do not try to commodify its dynamics. Nikolai's *Sick Trans-sex Gloria* is the closest to the actual world, but it is set in a post-apocalyptic science-fictional scenario (Figure 7.2). The remaining three pieces are radically abstract, from *Resurrection Lands*' trans-Black archive – consisting of a square surrounded by different build-ings, and various computer screens – to Qiang's *Vampyroteuthis Infernalis*, an abstract, unknown, dreamy space, where elements float freely. And Satterwhite's outer-space club, between utopia and uncanniness.

Nevertheless, the experience we have with these worlds is far from immersive. This is one of the main differences between the artworks and the mainstream metaverse. In *Resurrection Lands*, space is static. We can-not roam around the square as we stay in an overhead view until we click one of the pre-defined spaces, which transports us to a first-person video in which we do not control the view, or the position. In *Domestika*, we can decide where to look in the 360° space, but we cannot choose where to

FIGURE 7.2 Screenshot of *Sick Trans-sex Gloria* by Tabitha Nikolai.

Source: Courtesy of the artist.

move as the camera is pre-defined by the artist. Consequently, its maximalism loses disorientating abilities (Misra, 2020), as paths are strictly pre-codified.

In *Sick Trans-sex Gloria*, *Vampyroteuthis Infernalis* and *Sin Sol/No Sun*, we can control both the movement and the camera. In *Sin Sol/No Sun*, because of its XR technique. In *Sick Trans-sex Gloria*, through its closed environment, which results particularly small. Meanwhile, *Vampyroteuthis Infernalis* is a radically open free world composed of different environments that we can reach through portals. No decisions regarding where to go or where to look are pre-codified by the artist. Consequently, it is the sampled piece that better suits Ahmed's phenomenology of queer disorientation (2006).

7.3.4 Avatar and Identity

The analysed works are far away from metaverse's possibilities of avatar customization, identity construction, and embodiment. Only one, Qiang's *Vampyroteuthis Infernalis*, allows customization thanks to the affordances of the software it is stored in (Mozilla Hubs). Interestingly, its avatar possibilities are numerous, ranging from pre-defined to made-from-zero avatars, and from human to no-human embodiments, thus opening the gates to queer approaches to body and identity (Lanier, 2022).

Two of the pieces address our real identity. cárdenas' *Sin Sol/No Sun* signals our present as the last opportunity to prevent climate change. By the end of the piece, Aura addresses us directly: "REJECT YOUR PROGRAMMING/INTERRUPT ALGORITHMS OF DAILY RITUAL/ THIS IS NOT A NORMAL RUNTIME ENVIRONMENT/we must end this so-called order". Brathwaite-Shirley's *Resurrection Lands* is presented differently depending on our identity. This is paradigmatic in all her work. On the one hand, it is a demand for white and cis accountability. On the other, a celebration of Black love, care, and memory. Honesty and accountability are, therefore, fundamental, and avatar customization is forbidden. To define our identity, we must respond to two questions by the beginning of the piece: 1) "What team do you want to enter as?", with two possible responses: "Pro Black Pro Trans Team. A team composed of all Black trans people who want to meet their ancestors" or "Consumers Team. A team composed of those who are not trans"; and 2) "Terms and conditions. You must agree to support Black trans people to reap the rewards of being in their presence". If declined, the viewer is forbidden to enter the *Resurrection Lands*.

In the two remaining sampled pieces, different avatars/embodiments are deployed. In Satterwhite's *Domestika*, technical limitations impede any embodiment. Nonetheless, both Black identity and Black bodies are fundamental elements of Satterwhite's work, as we can see here in the recorded version of him and other queer bodies dancing half-naked. *Sick Trans-sex Gloria* supports a better sense of embodiment because we can walk the space and change the point of view. We soon realise that we are Gloria, the trans woman from the title. In any case, we never see Sonia's body, not even her arms and hands, which deepens the work's feeling of solitude.

7.3.5 Interactivity

In interactivity, the analysed pieces fail to reach immersive standards. Satterwhite *Domestika*'s interactivity, as stated previously, is limited to deciding where to look. Additionally, this interaction does not change the plot dramatically. In *Sick Trans-sex Gloria*, we can also decide where to go, and the storytelling is determined by this movement. While roaming through the building, we should look for different ways to reach the next floor, as on each of the floors a new message is hidden. In any case, no other interactivity is possible. *Sin Sol/No Sun* is, expectedly, more interactive due to its XR technique. By moving our phones, we can discover different space elements. At the same time, we can walk through this space. In any case, this interactive element is not particularly related to its storytelling. Only if played in a park does the piece's reference to the remaining trees make sense, as we see the real contemporary forests that should be preserved.

In Danielle Brathwaite-Shirley's *Resurrection Lands* the interaction is reduced to some clicks on the screen. However, these interactive clicks are fundamental to the plot. As stated previously, when entering the *Resurrection Lands*, two questions regarding our identity, the identity of our ancestors, and our political commitment to trans and Black rights will determine the entire experience of the piece. Regardless of the simplified mechanism of this interaction, its implications in thinking about how virtual reality and the metaverse can hold accountability and mobilise political discourses are remarkable.

The most interactive piece of the sample is Qiang's *Vampyroteuthis Infernalis*, due to the Mozilla Hub's affordances. This includes the possibility of text and voice chat with other spectators, thus becoming the only social work of art from the sample and, therefore, a world-making that allows community building. Moreover, some of its objects can be duplicated, grabbed, and changed; opening its world-making to the audience's willingness to render it differently. Lastly, it also allows us to introduce new elements into the composition, both predefined by the software or our creations.

7.4 CONCLUSION

When *Frieze*'s editor-in-chief Andrew Durbin suggested that artists, rather than Silicon Valley's developers, were the ones imagining the most interesting digital realms, the possibility of an alternative genealogy to the metaverse emerged. In this chapter, I have tried to offer a partial glimpse of that genealogy, one particularly focused on queer, BIPOC artists.

Metaverses and virtual realities are both "symptomatic" – mirroring the context they emerge in – and "productive" – actively perpetuating or modifying that context (Shaviro, 2010). Symptomatic and productive are complexly entangled, and there is not a pure symptomatic nor a unique productive proposal. On the one hand, Rocha and Snelting's concepts of "volumetric regimes" and the "industrial continuum of the metaverse" (2022) help us conceptualise mainstream metaverses such as Meta's as a symptomatic, conservative-oriented productive force. This is proved by the denounces of queerphobia, misogyny, and racism in their virtual environments (Leonard, 2020; Morla, 2022), and the ableism, cis-straight-normativity, and white supremacism of their avatar creator tools (Rocha & Snelting, 2022, pp. 191–198).

On the other hand, in the analysed artworks, we find change-oriented productive metaverses. However, their affordances are particularly limited. To create an alternative to Meta, we can conclude, queer artists should be more ambitious. The analysis shows that the pieces mix immersive and non-immersive resources. They put special effort into visual means – first-person point of view, single frame, three-dimensional, dynamic compositions – but they opt for non-immersive affordances when more difficult techniques are needed, especially in interactive elements or avatar customization. Of course, it was not their goal to build a holistic alternative to Meta. What we can conclude here is that, however, we lack queer-oriented ideas of interaction in the metaverse.

It is in aesthetics, storytelling, and world-making where we found their most original contributions to thinking the metaverse otherwise. Altogether, they share an inclination toward non-realistic depictions – from science fictional to abstract or conceptual – with a particular political position against hyper-realism. They do so by going back to previous video game generations' aesthetics, glitch politics, or thematic texturing. They also coincide in mixing rendering techniques (both recorded and modelled). Regarding content, rather than social networking, the analysed queer metaverses invest in compromise storytelling, with topics related to social justice, anti-racism, climate change, body, and identity politics, all of them approached through a queer intersectional approach. Last, in their world-making proposals, we find an alternative to Meta's attempt to mirror reality (such as the Tour Eiffel or the Sagrada Familia). Following Muñoz's (2009) call for queer imagination, these artists abandon real spaces and invest their rendering in imagined words, from science fiction (*Sick Transsex Gloria, Sin Sol/No Sun*) to abstraction (*Domestika, Vampyroteuthis Infernalis*), or in between (*Resurrection Lands*).

REFERENCES

Ahmed, S. (2006). *Queer phenomenology. orientations, objects, others*. Duke University Press.

Asante, G., Watley, E., LeMaster, L., & Johnson, A. (2021). Queer conversation with Amber Johnson and Lore/tta LeMaster. *QED: A Journal in GLBTQ Worldmaking*, 8(1), 174–184. https://doi.org/10.14321/qed.8.1.0174

Bennett, A., & Beckwith, M. (2018). Queering virtual reality with drag realness: A case study of a creative investigation. *Refractory Journal of Entertainment Media*, 30. https://refractoryjournal.net/30-bennett-beckwith/

Bizzocchi, J., & Tanenbaum, J. (2011). Well read: Applying close reading techniques to gameplay experiences. In D. Davison (Ed.), *Well played 3.0: Video games, value and meaning* (pp. 262–290). ETC Press.

Blanco-Fernández, V. (2022). Rendering volumetrically, rendering queerly. *A Peer Review Journal About, 11*(1), 104–115. https://doi.org/10.7146/aprja. v11i1.134308

Blas, Z. (2022). *Unknown ideals.* Sternberg Press.

Brummett, B. (2018). *Techniques of close reading.* Sage Publications.

Calleja, G. (2011). *In-game: From immersion to incorporation.* The MIT Press. https://doi.org/10.7551/mitpress/8429.001.0001

cárdenas, m. (2010). Becoming dragon: A transversal technology study. *Ctheory. Code Drift: Essays in Critical Digital Studies.* https://journals.uvic.ca/index. php/ctheory/article/view/14680

Carrubba, L. (2023). *Los videojuegos en disputa: la experiencia videolúdica y los juegos que vendrán.* Aguaderramada.

Casalini, G. (2021). Trans Ecologies of Resistance in Digital (after)Lives: micha cárdenas' Sin Sol/No Sun. *Media-N. The Journal of the New Media Caucus, 17*(2), 8–26. https://doi.org/10.21900/j.median.v17i2.769

D'Ignazio, C., & Klein, L. (2020). *Data feminism.* The MIT Press.

Durbin, A. (2022). Editor's letter. *Frieze, 231,* 13.

Galloway, A. (2006). *Gaming. Essays on algorithmic culture.* University of Minnesota Press.

Halberstam, J. (2011). *The queer art of failure.* Duke University Press.

Harris, J. U. (2021). A feeling of healing: Jacolby Satterwhite's kaleidoscopic vision of queer self care. In E. Chodos & A. Durbin (Eds.), *Jacolby Satterwhite: How lovely is me being as I am* (pp. 6–9). Miller Institute for Contemporary Art.

Hart, T. (2020, August 10). Dining on trauma: Danielle Brathwaite-Shirley talks trans-tourism, motherhood, & being a 'Freaky Friday everyday'. *AQNB.* www.aqnb.com/2020/08/10/dining-on-trauma-danielle-brathwaite-shirley-on-trans-tourism-motherhood-and-being-a-freaky-friday-everyday/

Hinds, J., Williams, E. J., & Joinson, A. N. (2020). "It wouldn't happen to me": Privacy concerns and perspectives following the Cambridge Analytica scandal. *International Journal of Human-Computer Studies, 143.* https://doi. org/10.1016/j.ijhcs.2020.102498

Isaak, J., & Hanna, M. J. (2018). User data privacy: Facebook, Cambridge analytica, and privacy protection. *Computer, 51*(8), 56–59. https://doi.org/10.1109/ MC.2018.3191268

Khanna, S., & Brathwaite-Shirley, D. (2021, April). Interview with Danielle Brathwaite-Shirley by flatness for feminist review and women's art library. *Feminist Review, 129*(1), 109–122. https://doi.org/10.1177/01417789211037031

Kouri-Towe, N. (2014). Queer apocalypse. Survivalism and queer life at the end. In S. Carte, (Ed.), *You cannot kill what is already dead* (pp. 17–20). University of Toronto Scarborough.

Lanier, A. (2022). *The rendered body: Queer Utopian thinking in digital embodiments.* Master Final Thesis, Massachusetts Institute of Technology.

Leonard, D. J. (2020). Virtual anti-racism: Pleasure, catharsis, and hope in Mafia III and Watch Dogs 2. *Humanity & Society, 44*(1), 111–130. https://doi.org/10.1177/0160597619835863

Mehta, I. (2022, August 16). Meta launches Horizon Worlds in France and Spain. *TechCrunch.* https://techcrunch.com/2022/08/16/meta-launches-horizon-worlds-in-france-and-spain/

Misra, S. (2020). Queer VR: Orientation and temporality in Jacolby Satterwhite's Domestika. *Spectator, 40*(2), 40–44.

Morla, J. (2022, January 7). Abusos sexuales, metaversos y realidades virtuales. *El País.* https://elpais.com/babelia/2022-01-07/abusos-sexuales-metaversos-y-realidades-virtuales.html

Muñoz, J. E. (2009). *Cruising Utopia. The then and there of queer futurity.* NYU Press.

Munster, A. (2006). *Materializing new media. Embodiment in information aesthetics.* The University of Chicago Press.

Nyong'o, T. (2019) *Afro-fabulations: The queer drama of Black life.* New York University Press.

Peters, J. (2022, October 11). Meta figured out legs for its avatars. *The Verge.* www.theverge.com/2022/10/11/23390503/meta-quest-horizon-avatars-legs

Rezaire, T. (2022). *Womb consciousness.* Les Presses du Réel.

Rocha, J., & Snelting, F. (2022). *Volumetric Regimes: material cultures of quantified presence.* Open Humanities Press.

Ruberg, B., & Shaw, A. (2017). *Queer game studies.* University of Minnesota Press.

Russell, L. (2020). *Glitch feminism. A Manifesto.* Durnell Marston.

Shaviro, S. (2010). *Post-cinematic affect.* O-Books.

Sivanesen, S. (2017). Queering and quaring virtual space. *Runway Journal, 35.*

Steyerl, H. (2011). In free fall: A thought experiment on vertical perspective. *E-flux Journal, 24.*

Stokel-Walker, C. (2022). Welcome to the metaverse. *NewScientist, 253*(3368), 39–43. https://doi.org/10.1016/S0262-4079(22)00018-5

Stone, A. R. (1991). Will the real body please stand up? In M. Benedikt (Ed.), *Cyberspace: First Steps* (pp. 81–118). The MIT Press.

Suri, H. (2011). Purposeful sampling in qualitative research synthesis. *Qualitative Research Journal, 11*(2), 63–75. https://doi.org/10.3316/QRJ1102063

Tavinor, G. (2022). *The aesthetics of virtual reality.* Routledge.

Thomas, S. A. (2022). Danielle Brathwaite-Shirley. *Frieze, 231,* 68–77.

Transmediale (2022, September 11). Vilém Flusser Resident 2020: Pete Jiadong Qiang. *Transmediale.* https://archive.transmediale.de/content/vil-m-flusser-resident-2020-pete-jiadong-qiang

Vaidhyanathan, S. (2021). *Anti-social media. How Facebook disconnects us and undermines democracy.* Oxford University Press.

CHAPTER 8

Film Practices in the Metaverse

Methodological Approach for Prosocial VR Storytelling Creation

Dr. Francisco-Julián Martínez-Cano

Universidad Miguel Hernández, Elche, Spain

Dr. Richard Lachman

Toronto Metropolitan University, Canada

Dhruva Patil

KLE, Dr. M.S. Sheshgiri College of Engineering and Technology, India

CONTENTS

DOI: 10.1201/9781003379119-8

8.1 INTRODUCTION

Although it sounds like a neologism of our times, the term "metaverse" was in fact originally coined by Neal Stephenson in *Snow Crash* (1992), a novel in which "some of the action takes place in the real world, but much of it takes place in a mass-visited communal virtual world called the Metaverse" (Taylor, 1997, p. 177). Stephenson's novel draws on earlier works like *Pygmalion's Spectacles* (Weinbaum, 1935), a story that describes a pair of virtual reality goggles with direct reference to movies in which the viewer has the power to participate in the story and to speak to the "shadows" who also talk back, breaking the barrier of the screen in this futuristic vision of what immersive VR movie experiences might one day be like. Other stories that could be identified as precursors to *Snow Crash* are *The Veldt* (Bradbury, 1950) and *The Trouble with Bubbles* (Dick, 1953). The literary tradition of the metaverse also includes Isaac Asimov's *The Naked Sun* (1956), William Gibson's *Neuromancer* (1984) and Ernest Cline's *Ready Player One* (2011). In the case of video games, a medium that connects more naturally with the concept of the metaverse, the idea of virtual worlds can be traced back to the text-based MUDs (multi-user dungeons) of the 1970s, which were followed by MUSHes (multi-user shared hallucinations) and also inspired video games such as *Habitat* (LucasFilm, 1986), *OnLive! Traveler* (Activeworlds, 1998) and *Second Life* (Linden Lab, 2003). *Second Life* was effectively the first online virtual world, with users embodied as avatars who could communicate and interact with other users in real time and with their own currency, a feature that offered the possibility of establishing markets on this virtual platform. This was followed in 2010 by *Minecraft*, whose user numbers surpassed 150 million in 2021. Other games offering virtual worlds are *Roblox* (Roblox Corp.) and *Fortnite* (Epic Games) (Ball, 2022, pp. 29–32), whose *Fortnite Creative* platform allows users to program events in its virtual environment, such as concerts, exhibitions and other cultural activities (Acevedo Nieto, 2022). For example, a DJ Marshmello concert in 2019 was attended by 10 million players simultaneously, while a Travis Scott concert in 2020 registered an audience of 12.3 million, according to data provided by Epic Games. Other artists have offered similar events, turning what began as a video game into a virtual online space for social interaction.

References to parallel virtual worlds in film can be found in the interpretation of the cyberspace of Gibson's *Neuromancer* in *The Matrix* (the Wachowskis, 1999–2021), in the film adaptation of the novel *Ready Player*

One (Spielberg, 2018) and in *Strange Days* (Bigelow, 1995). Other examples include *The Lawnmower Man* (Leonard, 1992), based on Stephen King's short story of the same name, and *Total Recall* (Verhoeven, 1990), based on the novella *We Can Remember It for You Wholesale* (Dick, 1966), offering dystopian visions that explore the potential of the technological development of VR for military training or for the creation and preservation of memories. All these examples demonstrate the existence of a connection between literature, film and video games in relation to the development of stories that reflect our interest in the creation of new non-places in VR and the metaverse. Indeed, these references are sometimes the most vivid for the general public, who may still struggle with understanding what, and indeed why, corporate investment in metaverse technologies could intersect with their day-to-day lives.

8.1.1 Defining the Metaverse

The metaverse and other virtual worlds are the product of an inherent human need to transcend and escape our non-virtual reality (Taylor, 1997, p. 179). In cyberpunk literature and cultural products, they have been associated with dystopias in which the society of the real world has collapsed due to the corruption of the government and other social powers. While it is currently still in the process of being defined, the conception of the metaverse in terms of technology and mass media associates it with the development and evolution of the internet as a capitalist socioeconomic construct (Cuevas-Hewitt, 2011) that is still the subject of debate. Some authors refer to it as the 3D internet, as the natural continuation or evolution of the World Wide Web. Mystakidis defines it as "the post-reality universe, a perpetual and persistent multiuser environment merging physical reality with digital virtuality" (2022, quoted in Linuo, 2022, p. 2). Another definition of the metaverse is the one proposed by Mathew Ball:

> A massively scaled and interoperable network of real-time rendered virtual 3D worlds that can be experienced synchronously and persistently by an unlimited number of users with an individual sense of presence, and with continuity of data, such as identity, history, entitlements, objects, communications, and payments.
>
> (Ball, 2022, p. 55)

Clearly, the metaverse is a realm of virtual environments in which individuals are able to interact and engage in experiences that complement and interconnect with the real world. It is thus "a distributing channel for narratives characterised by their degree of customisation and the level of interaction with other users" (Acevedo Nieto, 2022, p. 48). With such a potential to impact so many aspects of our online and offline life, it is important to ensure that the development of this channel does not result in the establishment of monopolies by private corporations or institutions. Instead, the same approach taken to the creation of the internet should be adopted, with universities, research centres and public institutions working on its foundations with the aim of preventing any attempts to centralise control over it. Although it is not yet possible to say exactly how it will be used in everyday life, whether it will be similar to how we use the internet of things, or whether parallel virtual worlds will bring new benefits to our lives, it is clear that it possesses a disruptive potential arising precisely from the unpredictable nature of its future. It is therefore important to focus efforts on the technological development and specific features of the medium and its content in order to contribute to its definition and to ensure that it is not subject to any form of manipulation by the political and/or economic powers that be, and that its objective is to improve conditions in contemporary societies. With this aim, the metaverse should be developed by independent users, developers, researchers and creators, and driven by small and medium-sized enterprises.

8.1.2 Virtual Reality Film and Filmmaking Practices in the Metaverse

The development of the metaverse and its immersive technologies is having an influence on contemporary filmmaking. Associated with the concept of the metaverse are the terms "VR film" (Chang, 2016), "VR cinema" (Zarka & Shah, 2016), "cinematic virtual reality" (Mateer, 2017) and "metaverse film" (Zhu, 2022, quoted in Linuo, 2022, p. 2). As new expressions of digital media, virtual reality (VR) and augmented reality have become hotbeds of pioneering experimentation with immersive audiovisual languages, resulting in the production of numerous narrative works by combining multiple layers of 360-degree video, 3D CGI and volumetric capture (Martínez-Cano & Roselló-Tormo, 2020), or creating XR experiences that combine the real and the virtual directly on AR devices, such as *Terminal 3* (Malik, 2019, quoted in Martínez-Cano, 2020). Other examples of productions that are breaking through media and format boundaries include those made in VR social environments, such as the VRChat app, which was used to make the documentary *We Met in Virtual Reality* (Joe

Hunting, 2022). Similarly, authors and filmmakers like Illya Szilak and Cyril Tsiboulski work together on VR audiovisual fiction in their series *Queerskins* (2019–present). With *Flesh and Sand* (*Carne y Arena*, 2017), Alejandro González Iñárritu experimented with immersive media to put viewers in the position of a migrant trying to cross the Sonora Desert in the hope of attaining the supposed American dream, while Ming-liang Tsai's *The Deserted* (2017), one of the longest VR films made to date, confronts the theme of death. Another example is *The Hangman at Home* (Michelle Kranot & Uri Kranot, 2020), which won the Grand Jury Prize for Best VR Immersive Work at the 77th Venice Film Festival in 2020 – Venice VR Expanded, which turns the spectator into an actor who interacts with others in the scenes. More recently, Benjamin Steiger Levine's *Marco & Polo* (2021), along with other VR films such as *Kusunda* (Felix Gaedtke & Gayatri Parameswaran, 2021), *Goliath: Playing with Reality* (Barry Gene Murphy & May Abdalla, 2021) and *Re-Educated* (Sam Wolson, 2021), were featured at Montreal's 2022 Phi Centre exhibition *Horizons VR*.

Recent years have seen a boom in immersive technologies, along with game engines for real-time rendering, algorithms and AI, encryption and blockchain systems with cryptocurrencies and NFTs, thanks to the development of new political structures with a major influence on the cultural sector. The revolution of the metaverse will move from culture to industry (and not the other way round, as happened with the internet), and film is emerging as one of its chief precursors:

> The impact of the formation of new cultural structures on art, aesthetics and popular culture is enormous, especially for the Metaverse media and film. Both of them have the qualities as the moving image, so they interpenetrate each other in more ways and at a faster pace.
>
> (Linuo, 2022, p. 4)

Technologies such as immersive VR and AR devices, motion capture sensors and volumetric video, photogrammetry, algorithms, big data, AI and machine learning, together with game engines and the increasing maturity and affordability of 5G technologies and graphics cards, will give rise to new artistic paradigms that will contribute to the development of the metaverse. As VR is one of the main technologies used for displaying moving images in the metaverse, the metaverse film is based on principles of immersiveness, participatory creation, embodiment and viewer presence in the scene and the action of the story itself, as well as the use of avatars and the possibility of the use of AI to create virtual film experiences that leave the passive viewer and the dictatorship of the movie theatre behind. The new medium will change the aesthetics of film and redefine the very nature of filmmaking itself (Martínez-Cano et al., 2022) and the industry's production and consumption models, giving rise to new audiovisual story-telling products in which "[t]he moving-image culture is being redefined once again" (Manovich, 2002, p. 308). In this new context, the language of interactivity will form an essential part of the metaverse film, whereby participation will decentralise the binary structure of the aesthetic relationship between creator and viewer.

Although immersive VR devices are not a prerequisite for the metaverse, their use can amplify and enhance the user experience through the combination of haptic devices and VR and AR visors (Smart et al., 2007; Dwivedi et al., 2022). The combination of these elements offers creative possibilities that can result in experiences in which the user is immersed in the space of the virtual world as a participant rather than a mere observer. Film and video game content creators and artists are therefore called upon to experiment with these media and to define the new audiovisual languages of the metaverse. This is why a keen awareness of the transformation of conventional filmmaking practices resulting from the new technologies and the evolution of audiovisual imaging itself, as well as the discursive and expressive strategies employed to engage the viewer with the story, are essential to the development of effective narratives, which can also function as prosocial tools.

This chapter considers filmmaking practice and its contribution to the development of the metaverse. The starting premise is that film serves a function of shaping its audience (Deleuze, 1987), and the possibilities offered by the new metamedium could enhance the impact of film content on viewers. These new immersive audiovisual storytelling productions could take up traditional cinema's function of transmitting values, enhanced by the new technological features and possibilities. With this in mind, this chapter describes the creation process for the first episode in the series *The Stigma Machine* (Martínez-Cano et al., 2022), a VR fiction film that aims to offer viewers a perspective-taking experience that can help to instil positive social values. On the basis of this case study, we propose an experimental methodology for the creation of VR experiences of this kind, combining the use of volumetric filming and 3D CGI using Unity, a cross-platform game engine.

8.2 THE STIGMA MACHINE

The Stigma Machine is an experimental VR fiction that aims to use the strategies of this medium as a means of fostering empathy and putting the viewer in another person's skin. Conceived of as a VR film, it addresses the issue of social stigma, using the virtual film medium as a prosocial tool that can raise audience awareness and foster supportive behaviours in relation to hot-button issues in contemporary society, such as tolerance of and respect for diversity in the LGBTIQA+ community. It is a VR film structured in four episodes, in which the viewer takes the protagonist's point of view and is immersed in some of his key experiences, related to his same-sex attraction in childhood, youth, adulthood and older years, from the perspective of the social stigma rooted in the heteronormative context. The viewer is thus placed in the position of the main character in four different situations, interacting in virtual contexts that can produce a personified immersive experience.

The aim of this project is to explore the new opportunities that VR technologies offer the contemporary media ecosystem, specifically storytelling and filmmaking practices. These technological advances provide ways for social actors to create innovative products and invite the public to participate in them and engage with them with the aim of creating a better society. Initiatives of this kind are especially important today, when the crisis of capitalism in a globalised world is resulting in polarisation in our societies.

8.3 PROPOSED METHODOLOGY

Based on a case study method, this chapter provides an overview of the production of an experimental short VR film, exploring Genette's concept of narrative focalisation as this concept has been redefined by François Jost and André Gaudreault. To understand what it means to view a story in our times, we need to consider the conventional filmmaking structures, techniques and systems placed at the service of the narration, while at the same time exploring the possibilities offered by VR technologies in terms of immersing the viewer in the audiovisual text. Narratology, focusing on visual perceptions (ocularisation) and aural perceptions (auricularisation), serves as a basis for analysing a short film that immerses viewers in a situation where they take the perspective of another person's reality, occupying that person's identity and personified experience. It is thus an embodied experience that results in the ubiquitous presence of the viewer inside the story, immersed in the filmic space.

The result is the first of four VR short films in the series *The Stigma Machine*, which viewers enter by occupying a space conceived of as a virtual installation that recreates the set for each of the acts that comprise this work. Viewers are positioned in the filmic space through the physicality of a location accessed using VR goggles that immerse them in the non-place of the fictional setting. They therefore access the story through their own gaze, in a way that breaks with any traditional conception of point of view in film. For the design of each scene, a VR storyboard was developed using Quill software for virtual illustration and an Oculus Rift visor, so that each scene can be visited as a virtual environment accessed via a QR code, as shown in Figure 8.1.

The image is recorded using volumetric filmmaking technology. Volumetric video creation is an increasingly popular trend in immersive content production (AR/VR/MR), characterised by interactive experiences created predominantly with 3D scanned images, through depth sensors such as Kinect 2 or Azure. Technologies such as volumetric video and photogrammetry are used with game engines like UDK and Unity to facilitate the viewer's ubiquitous presence in realistic environments reconstructed by recording real images in volumetric video format, combined with synthetic 3D elements, where the viewer's interaction affects how the events unfold.

A veritable video game–film hybrid, volumetric video production takes inspiration from related creative fields, such as the documentary,

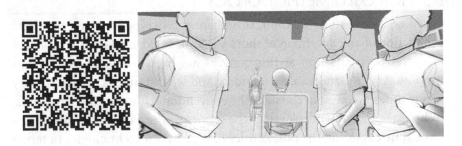

FIGURE 8.1   QR code for access to VR storytelling scene in the first episode of the series.

Source: Prepared by authors

immersive theatre, 360° VR video and interactive installations. This first episode was made using a Microsoft Kinect 2 sensor and 3D reconstruction of the spaces and elements that serve as settings for the action of the story. In simple terms, the resulting audiovisual experience consists of two layers:

- 3D reconstruction of the sets and elements of the space of action.

- Volumetric video recording with 4K cameras, Kinect sensors and Depthkit software to film the actors' performances on film sets with chroma backgrounds.

Apart from its technical and methodological conception, what makes this project innovative is its design as a prosocial tool intended to instil positive values in the audience. It therefore constitutes an important contribution not only to the activist mission of promoting tolerance and respect for diversity but also to experimentation with audiovisual language and its application in the context of filmmaking practices in the metaverse.

8.4 PRODUCTION PROCESS

The pre-production process was carried out in accordance with traditional film industry standards. A series of meetings were held with the group of child actors to work on ideas related to the staging of the scene. Notes were taken directly from these meetings to develop the final script. Meetings were held with the parents to sign the image rights and consent documents for their children's participation in the project. The filming was scheduled to take place over two days, on the 13th and 20th of May 2022, at the facilities of Miguel Hernández University of Elche in Alicante, Spain, using a Blackmagic 4k camera and a Microsoft Kinect 2 sensor connected to a Lenovo Legion Y720 computer with Depthkit 0.5.12 software.

The actors' performances were filmed over two days on a set with a chroma background. The furniture and props did not need to be constructed for the scene, as these were created later and inserted as 3D objects. Similarly, the classroom set was constructed on Autodesk Maya as a 3D element that the user can navigate around visually as the story unfolds. The performances were filmed first with the whole group and then with each actor individually, so that later each one could be inserted into the virtual environment in order to establish the right composition with the aim of conveying the intended meaning and eliciting the feeling we were aiming for from the audience. The sensor was positioned at the eye level of the character occupied by the user during the story, that is, the position of the child who is bullied by his classmates, as shown in Figure 8.2.

Using the Depthkit software, we set a resolution of 1080p and a depth-resolution of 512 × 424 with the Microsoft Kinect 2 sensor. While filming, we could turn our point of view to check the 180-degree volumetric 3D video recording, as shown in Figure 8.3.

FIGURE 8.2 Volumetric video recording process for the VR experience *The Stigma Machine*.

Source: Prepared by authors

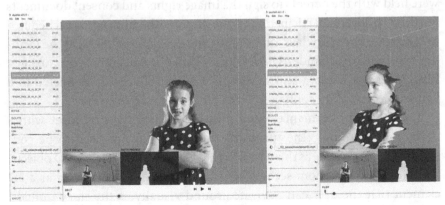

FIGURE 8.3 Screenshots during the volumetric video recording using Depthkit software.

Source: Prepared by authors

Once the recording was completed, we moved onto the post-production stage of refining the video clips using the editing panel. With the "isolate" option, we were able to set the two depth-of-field shots that would mark the beginning and end of our recording for export, and use Adobe After Effects' Keylight plug-in to eliminate the chroma background. Finally, we exported the mask generated using Adobe Media Encoder, in H.264 format on an mp4 file, maintaining the "Match Source" option on the preset parameter. Back in Depthkit, we performed additional mask refinement until we had a clean silhouette of each actor's clip. We exported our files, using the combined-per-pixel format optimized for using the Depthkit plug-in in the Unity game engine.

The post-production stage was carried out at the facilities of The Catalyst in the Creative School at Toronto Metropolitan University, using a PowerPC, Unity, an Oculus and an HTC Vive Pro headsets. For the individual creation of each child/character in the scene, the main challenge was to establish an idle video fragment that would keep the character active in the environment while other cast members are performing.

Since the children acted out their respective parts sequentially, the question of character activity was key to keeping the scene, as a whole, alive while each one is implemented individually. For this reason, each character was structured around an empty Unity object, which each of the dialogue clips and the clip in idle status were imported into as Depthkit clips. These clips stored in the project had to be opened and to have their parameters set. The metadata file produced by Depthkit on export had to be added, and at the same time the video clip had to be included on VideoPlayer, removing the awake player and placing the relevant. wav file in the audio. In addition, a Playable Director component had to be added in the same left column where we had to select a timeline that was also generated as a Unity object inside the scene, as shown in Figure 8.4. This allowed us to create a "video script playable track" on the timeline, where we could add a video clip player.

We imported and synched audio tracks, with each character generated. The same positions (x, y, z) had to be applied in the parameters for all Depthkit video clip objects that we had placed in the scene to avoid continuity errors related to the position of the actors in the virtual space. This process had to be repeated until the scene composition was complete; in this case, the characters were generated from 5 Depthkit volumetric video clips. Finally, a series of animation tracks were added to the

FIGURE 8.4   Screenshot during post-production and editing process in Unity.

Source: Prepared by authors

timeline to include the subtitles, which were generated as text objects using TextMeshPro. The solar system that incorporates the transition from day to night in the scene illumination, the fade-in and fade-out effects in the timeline and the home menu and all of its interactions were programmed with C# object-oriented language. Specifically, it was taken into account that the orientation of the solar system in relation to the 3D space of the classroom had to match the lighting of the characters in the Depthkit volumetric video clips.

Finally, once this VR experience is already completed, our aim is to conduct a preliminary study on the success of this content in achieving prosocial aims, as compared to more traditional 2D and non-interactive cinema, as well as how the pre-production and production methodologies of traditional film adapt to the process.

8.5 DISCUSSION AND CONCLUSIONS

We use the concept of a "metaverse film" to include all filmmaking practices related to the metaverse, including videos made inside virtual worlds like VRChat. In a sense Machinima, or films made over the last 15 years using game engines, could therefore be described as the origin of metaverse films. Similarly, immersive VR audiovisual storytelling forms part of filmmaking practices in this emerging metamedium, immersing the viewer in parallel virtual universes, and turning the audience into part of the story, inside the set itself. The term "metaverse film" could therefore be used for immersive audiovisual productions of this kind, as they generate new virtual universes and participate in the construction of the metaverse. As it is a current debate in the field, defining the metaverse as not as a single virtual environment but as the sum total of multiple metamedia universes impacts this conversation. VR has become a medium for the creation of immersive experiences where the user can occupy the position of a character inside the narrative diegesis, or simply participate as a viewer, but inside the filmic space. This allows for the construction of potentially more powerful discourses, with greater impact, identification and critical reflection on audiences.

The metaverse film does not seem poised to replace any of its predecessors in more traditional cinema, but it does offer new paradigms for storytelling through a hybrid mixture of traditional filmmaking methods and video game productions. The video game is clearly the natural reference medium for the use of VR headsets and, in general, the construction of interactive and immersive audiovisual narratives, perhaps because its target audiences demand a type of entertainment product in which they can participate and interact, regardless of whether they have the power to change the actual course of events in the story they are viewing. It is not only the filmmaking equipment that is updated and redesigned in these productions but also the models for enjoying the immersive audiovisual experiences of the metamedium. The metaverse transcends the two-dimensionality of the web as we know it. The metaverse film does the same with the two-dimensionality of the moving image and transforms the way we relate spatially to audiovisual fiction.

VR filmmaking practices contribute to the construction of the metaverse and the definition of audiovisual content that will be produced and consumed virtually. The production model and the method used to create the first episode of *The Stigma Machine* (Figure 8.5) is viable for the development of other immersive audiovisual stories that also offer a reflection on social conflicts and needs in an effort to instil positive values and a sense of justice in audience members. However, the use and application of VR films as prosocial tools needs to be approached with caution (Sora-Domenjó, 2022), as its medium- and long-term impact has not yet been demonstrated to be homogeneous for all potential users/viewers. Nevertheless, it is clearly better to develop new audiovisual products in the metaverse with these prosocial objectives rather than with other purposes or from other perspectives.

FIGURE 8.5   2D screenshots of the VR experience *The Stigma Machine*.

Source: Prepared by authors

FUNDING

This work is part of the research project: *The role of virtual reality audiovisual narratives in social inclusion and the perspective of prosocial models: Analysis of their characteristics, effects and impact on young university students*. Supported and funded by a grant from the Spanish Ministry of Universities under the State Program for the Promotion of Talent and its Employability in R+D+I, State Mobility Subprogram, of the State Plan for Scientific and Technical Research and Innovation 2017–2020.

The Stigma Machine series continues within the project *Immersive prosocial audiovisual narratives: Measuring their impact on society and analyzing their formal and technological characteristics*. Regional Ministry for Innovation, Universities, Science and the Digital Society under the Program for the Promotion of Scientific Research, Technological Development and Innovation in the Valencian Community – AICO 2022.

REFERENCES

Acevedo Nieto, J. (2022). Una introducción al metaverso: Conceptualización y alcance de un nuevo universo online. *adComunica*, 41–56. https://doi.org/10.6035/adcomunica.6544

Ball, M. (2022). *The Metaverse and how it will revolutionize everything*. Liveright Publishing.

Chang, W. (2016). Virtual reality filmmaking methodology (animation producing). *Techart: Journal of Arts and Imaging Science, 3*(3), 23–26. http://dx.doi.org/10.15323/techart.2016.08.3.3.23

Cuevas-Hewitt, M. (2011). Towards a futurology of the present: notes on writing, movement, and time. *The Journal of Aesthetics and Protest*, (8). www.joaap.org/issue8/finals/Cuevas-Hewitt,%20%20FINAL-WEB.doc

Deleuze, G. (1987). *La imagen-tiempo* (Vol. 2, p. 229). Paidós.

Dwivedi, Y. K., Hughes, L., Baabdullah, A. M., Ribeiro-Navarrete, S., Giannakis, M., Al-Debei, M. M., . . . & Wamba, S. F. (2022). Metaverse beyond the hype: Multidisciplinary perspectives on emerging challenges, opportunities, and agenda for research, practice and policy. *International Journal of Information Management, 66*, 102542. https://doi.org/10.1016/j.ijinfomgt.2022.102542

Linuo, Z. (2022). What's Metaverse film? Sci-fi, DAO or Digital installation?. *Revista FAMECOS, 29*(1), e43354. https://doi.org/10.15448/1980-3729.2022.1.43354

Manovich, L. (2002). *The language of new media*. MIT press.

Martínez-Cano, F. J. (2020). Volumetric filmmaking, new mediums and formats for digital audiovisual storytelling. *Avanca| Cinema*, 606–614. https://doi.org/10.37390/avancacinema.2020.a168

Martínez-Cano, F. J., Pérez-Caballero, J. M. L. A., & Hernández-Martínez, E. (2022). Cine de realidad aumentada: Reformulación del aparato cinematográfico: Estudio de caso de a Jester's Tale. *Fonseca, Journal of Communication*, (24), 109–125. https://doi.org/10.14201/fjc.28303

Martínez-Cano, F. J. & Roselló-Tormo, E. (2020). La dirección y realización audiovisual de realidad virtual. Análisis de Queerskins: A Love Story, una aproximación al cine volumétrico. *ASRI: Arte y sociedad. Revista de investigación*, (18), 111–125. www.eumed.net/rev/asri/18/realidad-virtual.html

Mateer, J. (2017). Directing for cinematic virtual reality: How the traditional film director's craft applies to immersive environments and notions of presence. *Journal of Media Practice, 18*(1), 14–25. https://doi.org/10.1080/14682753.2017.1305838

Mystakidis S. (2022). Metaverse. *Encyclopedia, 2*(1), 486–497. https://doi.org/10.3390/encyclopedia2010031

Smart, J., Cascio, J., Paffendorf, J., Bridges, C., Hummel, J., Hursthouse, J., & Moss, R. (2007). A cross-industry public foresight project. In *Proc. Metaverse Roadmap Pathways 3DWeb* (pp. 1–28). www.metaverseroadmap.org/MetaverseRoadmapOverview.pdf

Sora-Domenjó, C. (2022). Disrupting the "empathy machine": The power and perils of virtual reality in addressing social issues. *Frontiers in Psychology, 13*. https://doi.org/10.3389/fpsyg.2022.814565

Taylor, J. (1997), The emerging geographies of virtual worlds. *Geographical Review, 87*, 172–192. https://doi.org/10.1111/j.1931-0846.1997.tb00070.x

Zarka, O. M., & Shah, Z. J. (2016). Virtual Reality cinema: A study. *International Journal of Research and Analytical Reviews (IJRA), 3*(2), 62–66. http://ijrar.com/upload_issue/ijrar_issue_276.pdf

Zhu, J. (朱嘉明). (2022). *Metaverse and digital economy*(《元宇宙与数字经济》) (p. 290). China Translation & Publishing House.

CHAPTER 9

Video Game Design

A Blueprint for the Metaverse and Education

Dr. Kris Alexander

Toronto Metropolitan University, Canada

CONTENTS

9.1 HUMAN LEARNING AND PERSUASION

Humans can engage with information in three core ways, via audio, text, and video, and these are the core components, the semiotics of multimedia instructional design in traditional classroom education (Mayer, 2003). The core component of video games is the complex interweaving of those core ways of learning, in addition to added variables of affordances, and human autonomy, that is, what players can do, and when they can do it.

The affordances of interactive experiences refer to the options available for participants within a particular world, or digital environment. Gibson, in 1954, spoke about the relationship between motion and movement (as cited in Greeno, 1994), and in interactive experiences. For example, the function of 'jump', and 'when one can jump'. Within the context of

DOI: 10.1201/9781003379119-9

the aforementioned, 'motion' and 'jump' refer to the rules of a particular world, while movement and jumping are created by a world builder, providing participants with perceived autonomy. Perceived autonomy describes what video games and most interactive experiences are fundamentally about: persuasion.

Persuasion in interactive experiences is a direct subset of rules and affordances, in that they are precisely what a world builder *allows* its participants to do. To clarify, there is no single experience, digital or analog, that can allow participants to go everywhere and do everything. As such, persuasion is a tool used to *force* players into following the rules of the experience. Successful persuasion in interactive experiences has its participants accept the rules and affordances of the experience, much like what is seen in video games, but more explicitly in training and simulation. There are seven persuasive technologies which highlight specifically how players are guided through digital experiences: 1) reduction; 2) tunnelling; 3) tailoring; 4) suggestion; 5) self-monitoring; 6) surveillance; and 7) conditioning (Bogost, 2007).

Reduction refers to the simplification of player tasks in a way that removes extraneous information. For example, in most video games, walking is performed by pressing one button, or holding a joystick in a particular direction. This is a reduction because it does not assign one button per foot, as this would detract from the more pertinent story and/ or objectives. Tunnelling refers to a particular pathway or route that specifically places limitations via a designated set of waypoints or check-in spots. This can be a helpful tool for guiding players in the right direction and eliminating confusion that can lead to frustration. Suggestion refers to just-in-time information provided to participants at the most appropriate time of need. This is particularly beneficial for guiding players to the next story element and encouraging exploration and engagement with the environment. Tailoring refers to information presented to participants as a means of triggering a change in a particular behaviour. It is based on the idea of providing information that helps players in their current state or encourages them to explore something new. Self-monitoring refers to a participant's ability to see how they are doing within the context of a created world or experience. For example, providing feedback in the form of scores or rewards can be a way to show players how well they are doing, and what they need to do to progress. Surveillance refers to the ability for participants to compare themselves to another participant or a standard of behaviour based on the rules of the experience. For instance, leader boards or badges can help motivate a participant to work harder and improve their performance, as reflected by other elements in the interactive world. The designing of interactive experiences for educational purposes presents its own set of challenges within the realm of higher education. It can be difficult to ensure that game mechanics effectively support learning objectives while simultaneously engaging students (Lameras et al., 2017). In addition, principles of design thinking can be used to address the shortcomings of engagement and usability (Scholten & Granic, 2019). However, more research is required to fully comprehend the efficacy of these interventions and the optimal implementation strategies (Straker et al., 2015).

Both human learning and persuasion are core components of interactive world design, and these need to be considered when thinking about the metaverse, which will need to borrow from video games – the medium that has over two billion worldwide players who are accustomed to that culture and format of interactivity (Newzoo, 2020).

9.2 THE METAVERSE: ONLINE, COMMUNITY, AND SHARED EXPERIENCES

Neal Stephenson coined the term 'metaverse' in his 1992 novel *Snow Crash* to describe a virtual environment that combines the capabilities of the internet and the immersive qualities of virtual reality (Cheng et al., 2022). The metaverse allows users to interact with virtual objects and environments in real time, creating a seamless and immersive experience that blurs the line between the physical and digital worlds (Rospigliosi, 2022). Consequently, education, communication, commerce, and entertainment can be affected by the potential benefits of the metaverse (Gadalla et al., 2013). In education, the metaverse could provide students with immersive and interactive learning environments that facilitate learning (Kanematsu et al., 2014), and new opportunities for social interaction, cooperation, and problem-solving – essential skills for success in the modern world (Halamek, 2008). The metaverse currently consists of three core components: 1) an online space, 2) a community space, and 3) a place where humans participate in shared experiences (Breakfast Television, 2022).

The initial successful integration of online components of the metaverse appear in the 2003 experience *Second Life*, which is still seeing user engagement today (Boulos et al., 2007). As background, *Second Life* is a virtual world platform that allows users to not only create things but also construct virtual homes and travel through 3D environments virtually. Community members have also innovated on the platform by creating nuanced user groups, enabling them to participate in film production (Ziemsen, 2017). These online interactions are maintained not by the developers who created the worlds but by the communities that sustained the worlds via their individual participation and contribution.

Grand Theft Auto V is an effective example of the metaverse as a community space in which pedagogical components have been integrated into a professional practice, such as media production. In 2015, Rockstar released the Rockstar Editor, a platform that allowed players to record, edit, and share videos made from Story Mode and GTA Online footage (Rockstar Games, 2015). Here there exists a community of producers, directors, project managers, community managers, narrative designers, and many professions traditionally associated with the video game and esports industries (Kumu, 2020). There are films like *Overheat* that were created with the editor and have received over seven million views from community members (Vuko100, 2018). The members of these online spaces connect socially and professionally through shared objectives of film production. Examples such as *Grand Theft Auto V* are metaverse-like spaces in which individuals participate in experiences via shared objectives. Participant objectives are present in many video games, including Nintendo's *Animal Crossing: New Horizons* for social agriculture conversations (Nintendo, 2022), *Eleven Table Tennis* for virtual reality events and social gatherings (Eleven VR, 2022), and even Capcom's *Street Fighter 6* for global discussions about fighting and dance cultures (Capcom, 2023). In these situations, each participant establishes an accord between the constructed world, its affordances, behaviours, and its inhabitants.

As the core descriptions of the metaverse continue to formulate, the significance of the specificity of online space, community, and shared goals must remain at the forefront of the discussion; otherwise, the manifestation of the internet could resemble previous iterations of computer technologies, in which users are increasingly removed from the components and processes underlying designed experiences.

9.3 VIDEO GAME DESIGN AND THE METAVERSE

Considering the ways in which video games and interactive 3D spaces cater to human learning via rules and affordances, it is useful to explore the engaging features of video game design that exist separately from the entertaining elements. Prioritizing objectives will be required if the metaverse is to have a substantial social influence that extends beyond the last era of social media (Lanier, 2018). Given the current limits of computer hardware, internet latency, and even pedagogical content, education is likely to be a key topic in terms of accessibility. As such, the current focus of the metaverse is on breadth rather than depth, as seen in its graphical fidelity and, primarily, commercial manifestations (Meta, 2022). These experiences, as with many video games and interactive 3D spaces, are primarily didactic in nature, affording participants very little in terms of autonomous freedom. This is clear in advertisements and campaigns that focus on company intention rather than user attention.

The components of video games provide critical examples of successful forays into education that could help shape the metaverse, particularly when it comes to creating knowledge exportability, that is, the ability for users to add to their own lives outside of the experience itself.

Video games have a unique ability to transfer multiple messages simultaneously through interactivity. Magic Keys VR, for instance, is an application that enables users to utilize any available physical piano to view notes and learn how to play using tactile, aural, and visual feedback. In addition, the public can submit their own works and feed them into the application, making it a digitally forward-looking resource for piano instructors and musical training institutes seeking to expand their digital reach (Hackl, 2021). *Sokobond* is a video game for learning chemistry that uses puzzle strategies to captivate players while educating them about how components link in a relaxing environment (Hazelden & Lun, 2013). *Nanotale – Typing Chronicles* teaches players touch typing, the ability to type on a computer keyboard without looking at one's hands. In a fictional environment, players are tasked with gathering magical components that can only be attained through precisely scaffolded typing experiences (Cactus, 2021).

The goals of the metaverse will require careful consideration, as the current approach is a storefront model in which consumers take and remain contained. The academic world revolves around portable knowledge, which must remain central to the metaverse for it to have a substantial social impact (Clark et al., 2016). Experiences that emphasize self-sufficiency, long-term relationship building, and future-oriented skills such as computer programming could help the metaverse become an environment that provides more than it extracts from its users.

9.4 BENEFITS OF VIDEO GAME DESIGN

Focusing on the benefits of video game design is essential to the creation and evolution of the metaverse, as it is currently the medium that best connects simultaneously with human learning. Innovative technologies and features are often iteratively tested and refined in video games before being applied in other parts of the digital world, which can help inform future decisions about crafting thoughtful and engaging experiences for learners (Bardzell & Shankar, 2007). In terms of students and learning, for instance, students who participated in digital game-based learning had much higher levels of engagement and better learning outcomes than those who did not (Clark et al., 2016). In addition, addressing future skills, particularly those requiring computer technology, video games have been demonstrated to be an effective means of engaging kids in STEM education and fostering problem-solving skills (Khalili-Mahani & De Schutter, 2019; Li & Huang, 2021).

In addition to its educational benefits, the design of video games offers cognitive benefits for players, as they cater to aural, textual, visual, and interactive learning styles (Alexander, 2016), including how using an integrated perceptual-learning video game results in extensive visual advantages (Deveau et al., 2014). Further, a study protocol indicated that an acute intervention consisting of high-intensity physical exercise followed by a brain training video game had rapid impacts on the brain activity of older individuals during a Stroop task as measured by fMRI (Himmelmeier et al., 2019). Some video games can also be used to encourage perspective-taking and empathy among players (Dishon & Kafai, 2020). In the field of history and heritage education, video game design can be used to create immersive virtual reality experiences that increase students' awareness of the past (Egea-Vivancos & Arias-Ferrer, 2021; Hanes & Stone, 2019). There are video game design classes that explore corporate culture and workplace consent as part of the education surrounding game development, via titles like *Say No! More* by Thunderful Games (2021).

Video game design is crucial to the formation and growth of the metaverse. Students who engaged in digital game-based learning had significantly higher levels of engagement and superior learning results. Players can also be taught perspective-taking and empathy through video games. Focusing on the aspects of video games that transcend entertainment and the act of playing will be a crucial step in extracting the value of video games in interactive interactions in the metaverse.

9.5 CONCLUSION

Video games have evolved significantly since their early origins as simple forms of entertainment. Currently, they are acknowledged as a sophisticated medium that may support immersive and interactive learning, encourage social interaction and problem-solving abilities, and educate a broad variety of courses and skills. As such, they have the potential to play a crucial part in the creation of the metaverse, a term referring to permanent, rich, and sophisticated virtual worlds that can be accessible through a variety of modalities including audio, text, video, and interactivity. The metaverse has the potential to change education and learning by offering new immersive and interactive experiences that assist in the acquisition and transfer of information. It has the potential to establish a new form of communal space where individuals may engage in meaningful and relevant shared experiences and activities.

However, the precise form of the metaverse is yet unknown, since different theories exist regarding what it will be and what individuals will be able to accomplish within it. Some businesses envisage it as a marketplace, whilst others envision it as a decentralized arena with several interfaces. To comprehend the educational and learning potential of the metaverse, it is necessary to examine the intricacies and intended purposes of video games. By understanding the semiotics of the medium and the unique ways in which video games might be utilized for education, we can better contextualize the potential future of the metaverse and emphasize the benefits it can offer for classroom learning. The design of video games may inform the development of the metaverse as an educational tool and give significant insights into the problems and potential of implementing video game design concepts into educational situations. There is much that the metaverse education can learn from video games, particularly if the focus remains on the components of video games that extend beyond playing, focusing on how humans can acquire new information and skills.

REFERENCES

Alexander, K. B. (2016). *Video design and interactivity: The semiotics of multimedia in instructional design.* PhD, Concordia University https://spectrum. library.concordia.ca/id/eprint/981606/

Bardzell, S., & Shankar, K. (2007). Video game technologies and virtual design: A study of virtual design teams in a Metaverse. In R. Shumaker (Ed.), *Virtual reality* (Vol. 4563, pp. 607–616). Springer Berlin Heidelberg. https:// doi.org/10.1007/978-3-540-73335-5_65.

Bogost, I. (2007). *Persuasive games: The expressive power of videogames*. MIT Press.

Chen, Z. (2022). Exploring the application scenarios and issues facing Metaverse technology in education. *Interactive Learning Environments*, 1–13.

Cheng, R., Wu, N., Varvello, M., Chen, S., & Han, B. (2022). Are we ready for metaverse? A measurement study of social virtual reality platforms. In *Proceedings of the 22nd ACM internet measurement conference (IMC '22)*, October 25–27, 2022, Nice, France. ACM, New York, NY, USA, 15 pages. https://doi.org/10.1145/3517745.3561417

Clark, D. B., Tanner-Smith, E. E., & Killingsworth, S. S. (2016). Digital games, design, and learning: A systematic review and meta-analysis. *Review of Educational Research*, *86*(1), 79–122.

Deveau, J., Lovcik, G., & Seitz, A. R. (2014). Broad-based visual benefits from training with an integrated perceptual-learning video game. *Vision Research*, *99*, 134–140.

Dishon, G., & Kafai, Y. B. (2020). Making more of games: Cultivating perspective-taking through game design. *Computers & Education*, *148*, 103810.

Egea-Vivancos, A., & Arias-Ferrer, L. (2021). Principles for the design of a history and heritage game based on the evaluation of immersive virtual reality video games. *E-Learning and Digital Media*, *18*(4), 383–402.

Eleven VR. (2022). *Eleven VR* [online]. https://elevenvr.com/en/.

Gadalla, E., Keeling, K., & Abosag, I. (2013). Metaverse-retail service quality: A future framework for retail service quality in the 3D internet. *Journal of Marketing Management*, *29*(13–14), 1493–1517.

Greeno, J. G. (1994). Gibson's affordances. *Psychological Review*, *101*(2), 336–342. https://doi-org.ezproxy.lib.torontomu.ca/10.1037/0033-295X.101.2.336

Hackl, D. (2021, August 31). *Magic keys :: Mixed reality piano learning* [online]. https://magickeys.app/.

Halamek, L. P. (2008). The simulated delivery-room environment as the future modality for acquiring and maintaining skills in fetal and neonatal resuscitation. *Seminars in Fetal and Neonatal Medicine*, *13*(6), 448–453.

Hanes, L., & Stone, R. (2019). A model of heritage content to support the design and analysis of video games for history education. *Journal of Computers in Education*, *6*(4), 587–612.

Hazelden, A., & Lun, L. S. (2013. *Sokobond* [online]. www.sokobond.com/.

Himmelmeier, R. M., Nouchi, R., Saito, T., Burin, D., Wiltfang, J., & Kawashima, R. (2019). Study protocol: Does an acute intervention of high-intensity physical exercise followed by a brain training video game have immediate effects on brain activity of older people during stroop task in fMRI? – A randomized controlled trial with crossover design. *Frontiers in Aging Neuroscience*, *11*, 260.

Khalili-Mahani, N., & De Schutter, B. (2019). Affective game planning for health applications: Quantitative extension of gerontoludic design based on the appraisal theory of stress and coping. *JMIR Serious Games*, *7*(2), e13303.

Kirkcaldy, A. (2020). Video game industry statistics, trends and data in 2022 [online]. *WePC | Let's build your dream gaming PC*. www.wepc.com/news/video-game-statistics/.

Kumu, P. (2020). Gamebadges [online]. *Kumu*. https://kumu.io/gamebadges/gamebadges.

Lameras, P., Arnab, S., Dunwell, I., Stewart, C., Clarke, S., & Petridis, P. (2017). Essential features of serious games design in higher education: Linking learning attributes to game mechanics. *British Journal of Educational Technology*, *48*(4), 972–994.

Lanier, J. (2018). *Ten arguments for deleting your social media accounts right now*. Henry Holt and Company.

Lee, H., Woo, D., & Yu, S. (2022). Virtual reality metaverse system supplementing remote education methods: Based on aircraft maintenance simulation. *Applied Sciences*, *12*(5), 2667.

Mayer, R. E. (2003). The promise of multimedia learning: Using the same instructional design methods across different media. *Learning and Instruction*, *13*(2), 125–139.

Meta. (2022). *Meta | Social Metaverse Company* [online]. https://about.meta.com/.

Nintendo. (2022). *Animal crossing™: New horizons for the Nintendo Switch™ system – official site* [online]. https://animal-crossing.com/new-horizons/.

Rockstar Games. (2015). *Introducing the rockstar editor – rockstar games* [online]. www.rockstargames.com/newswire/article/4k41288381a711/introducing-the-rockstar-editor.

Rospigliosi, P. (2022). Adopting the metaverse for learning environments means more use of deep learning artificial intelligence: This presents challenges and problems. *Interactive Learning Environments*, *30*(9), 1573–1576.

Scholten, H., & Granic, I. (2019). Use of the principles of design thinking to address limitations of digital mental health interventions for youth: Viewpoint. *Journal of Medical Internet Research*, *21*(1), e11528.

Straker, L. M., Fenner, A. A., Howie, E. K., Feltz, D. L., Gray, C. M., Lu, A. S., Mueller, F. "Floyd", Simons, M., & Barnett, L. M. (2015). Efficient and effective change principles in active videogames. *Games for Health Journal*, *4*(1), 43–52.

Suzuki, S., Kanematsu, H., Barry, D. M., Ogawa, N., Yajima, K., Nakahira, K. T., Shirai, T., Kawaguchi, M., Kobayashi, T., & Yoshitake, M. (2020). Virtual experiments in metaverse and their applications to collaborative projects: The framework and its significance. *Procedia Computer Science*, *176*, 2125–2132.

Thunderful Games. (2021). *Say No! More*. https://thunderfulgames.com/games/say-no-more/

Vuko100. (2018). *GTA V Movie – OVERHEAT [4K]*. www.youtube.com/watch?v=_vhwRYhWAzk

Ziemsen, E. (2017). *Developing a learning model for teaching film production online*. University of British Columbia.

CHAPTER 10

Film Production and Architecture Education in the Metaverse

Dr. Eva Ziemsen

Humber College, Canada

Elizabeth Fenuta, OAA, M. Arch

Humber College, Canada

CONTENTS

10.1 INTRODUCTION

Film production and architecture are two fields that are uniquely poised for professional practice and education in the metaverse, given innovations in visualisation software, the increasing demand for immersive narratives,

DOI: 10.1201/9781003379119-10

design and training and the global shift to virtual collaboration. To explore the specifics of how film production, specifically virtual production, and architecture, and the education of these fields, is possible in the metaverse, it is prudent to provide a widely accepted understanding of the metaverse. According to Matthew Ball (2022), the metaverse is, "A massively scaled and interoperable network of real-time rendered 3D virtual worlds that can be experienced synchronously and persistently by an effectively unlimited number of users" (p. 29). While a robust vision exists, the actual metaverse is still being built and debated by tech giants, like Meta and Nvidia, and smaller developers alike, which has led to the formation of a Metaverse Standards Forum (The Metaverse Standards Forum, 2022). The education of film production and architecture in the metaverse will be profoundly impacted by the overall standards that inform ethics, safety, design, and technical execution.

10.2 EDUCATIONAL FRAMEWORK FOR FILM PRODUCTION AND ARCHITECTURE IN THE METAVERSE

Educators are accustomed to constant new technical disruptions that have the potential to impact their teaching methodologies. Applying a robust framework to the adoption of teaching film production and architecture in the metaverse could inform, early adoption enthusiasm and address hesitancies of educators of its practical application. One example of an applicable framework for assessing the adoption of new technology is called the "SECTIONS Model" which stands for, Students, Ease, Cost, Teaching and Learning, Interactivity, Organisational Issues, Novelty, and Speed (Bates & Poole, 2003) and provides a comprehensive assessment that addresses the perspectives of many stakeholders who oversee adoption of institutional educational technology. For example, in assessing organisational issues of adopting teaching film production or architecture in the metaverse, questions around infrastructure, employee training, and technical support would arise. Teaching and learning would be analysed from multiple angles, including a pedagogical grounding in learning theories that are foundational to educational technology. A theoretical grounding for harnessing metaverse technology is effectively aligned with the constructivist learning theory, which asserts that "learning is affected by the context in which an idea is taught" (Bada & Olusegun, 2015) and originated with a variety of works, including that of Jean Piaget, who stated, "In all behaviour patterns it seems evident to us that learning is a function of the environment" (Piaget, 1952). VR-based learning is aligned with a constructivist learning theory as it enables students to become fully immersed, with the use of VR headsets and other accessible interfaces, in the environments related to their learning. For example, in the case of films, students learning about virtual production can now virtually enter a VP volume/stage for any length of time (*METAVERSE X APG*, 2023), which would be a rare opportunity in the physical world for most students at this time (Ziemsen & Fenuta, 2022). In presenting examples of how the education of film and architecture education can be implemented in the metaverse, it is important to note the extremely common dismissals of immersive learning by sceptical stakeholders, sometimes educators who have reservations in each category of the "SECTIONS" model, (Bates & Poole, 2003), based on anecdotal, personal perspectives, and students, who may have had poor previous exposures to VR technology in education. The way new metaverse technology is introduced into the ecosystem of an educational setting is correlated to its effective adoption. An ecosystem metaphor works well for defining how a new technology is integrated into an educational institution (Ziemsen, 2017). For instance, Zhao and Frank (2003) compare the successful integration of technology into classrooms with the introduction of zebra mussels to the Great Lakes. As they show, a variety of intricate factors will play a role in how successfully a technology is adopted or developed within the framework of a school system (Zhao & Frank, 2003). Therefore, the adoption of teaching film production and architecture in the metaverse will depend on factors such as existing faculty members who will aid in the acceptance and survival of the 'new metaverse' approach to teaching and learning.

The real-world demands to innovate and push the education of practical subjects like film production and architecture into the metaverse are supported by survey studies of students, industry-specific reports on exponential growth and demand for personnel, and an overall technological evolution that is happening in all sectors. As concluded in a study examining "Student Thoughts on Virtual Reality in Higher Education" (Cicek et al., 2021), students are calling for VR integration and alternatives to the traditional education and learning processes (Cicek et al., 2021). An emphasis on teaching 21st century skills, the pressure to reduce school operational expenditures and rising enrolment by students needing flexible learning (Truong, 2022) will lead to institutions being more amenable to the innovation of online film production courses. The worldwide virtual production industry will increase to 5.4 bn by 2026 (Infogence Global

Research, 2021) and will demand graduates with relevant skills in virtual production with agility to adapt to constantly evolving technology. In the educational setting, students will become more inclined to use virtual production as a pipeline for creative expression because of the democratisation of access and open resources and with impetus to join the film industry's biggest gold rush since the conversion from celluloid to digital video. In an architectural context, designers and homeowners alike will have access to tools that revolutionise the concept of architectural representation. "By 2032, each and every one of us will have the opportunity to easily build our own Metaverse homes and fill them with personal and professional tools" (Terry et al., 2002, p. 9).

10.3 MACHINIMA, VIRTUAL PRODUCTION, AND CONTENT CREATION IN THE METAVERSE

To understand the vast potential of virtual production in the metaverse, it is important to situate the present moment in the history of virtual filmmaking. The concept of producing and teaching film production in a virtual world is not new (Ziemsen, 2017) and stems from the method of 'machinima', which is, "animated filmmaking within a real-time virtual 3D environment" (Marino, 2004, p. 1). One very famous example using machinima was the show *Red vs. Blue* by the company Rooster Teeth, who filmed within the game *Halo*, owned by Microsoft, who later hired the team to produce more content (Johnson & Pettit, 2012, p. 187). These grassroots of producing film content 'in-game' gave way to virtual production (Harwood & Grussi, 2021). Virtual production is an evolved pipeline of film production that involves harnessing the power of a game engine, predominantly Epic Games' Unreal Engine (UE), to create content in two pipeline approaches, ICVFX and 'in-engine'. ICVFX refers to in-camera visual effects, requiring creators to have access to a studio-based LED volume, using motion tracking technology, to film live performers and set pieces in the foreground, with virtual environments displayed on the LED wall in the background, to create a real-time motion picture, seamlessly combining physical and virtual elements in one shot. In-engine productions, also sometimes referred to as real-time animation, do not require physical space and are produced solely in-engine. Here, graphics and realism are improving in quality at a rapid rate, and major Hollywood Studios are committing to this pipeline for their future slates. For example, Spire Animation Studios announced a 20-million-dollar deal with Unreal Engine that would, "enable Spire to produce the highest quality animated visual content . . . while simultaneously building out worlds and experiences for the metaverse" (Goldsmith, 2022). Notably, these studios recognize the exponential potential in producing content via virtual production for multi-platform delivery, including metaverse storyliving. Disney has already trademarked the term storyliving, specifically, "Storyliving by Disney"™ (Storyliving™ by Disney, 2022) and is revealing its plans to "connect the physical and digital worlds . . . allowing for storytelling without boundaries" (Milmo, 2021).

Currently, in-engine pipelines are completed using the software interface of Unreal Engine or Unity; however, in the not-so-distant future, creators will circle back to a machinima-type workflow, whereby you would don a VR headset and simply teleport to any virtual environment setting to start filming (Ziemsen & Fenuta, 2022). This may seem far-fetched and unnecessary to some who are currently adapting to the rapidly evolving VP pipeline in the physical world; however, anyone who has been here from the beginning understands this is the future of film production. An early adopter, George Bloom, outlined an early understanding of the future of film production in his Tedx Talk, "Virtual Reality – How the Metaverse will change Filmmaking" (TEDx Talks, 2014). Bloom detailed components of the disruptive technology that would enable filmmaking in the metaverse, including photogrammetry, the process of scanning real-world objects to create 3D assets, and digital humans/avatars, via motion capture technology (TEDx Talks, 2014). At the time of his talk, this was certainly new and revolutionary thinking; however, now it is being done by most virtual production companies operational today. The metaverse will provide a setting in which film production can occur in a professional and educational context. Film/media producers and educators will play a central role in creating the narratives *in and of* the metaverse.

Central to many metaverse narratives is the game engine of Unreal Engine, which is part of a larger and highly relevant ecosystem of the company Epic Games, including ancillary products, assets, plug-ins, and existing examples of the metaverse. Foremost, Unreal Engine itself has historically been known as a game engine but is now being used as a filmmaking platform and tool for virtual production. No matter which pipeline approach is pursued, ICVFX or in-engine, a virtual asset is required and is often easily obtained pre-made in UE Marketplace, which offers users thousands of optimised environments, objects, characters, animations, and textures/materials (*Unreal Engine Marketplace | Store of UE Assets for Games and 3D Rendering*, n.d.). While larger companies produce their own original assets, UE offers a generous IP policy that enables indie and student users to use assets that would normally costs thousands of dollars for relatively no significant cost. Since all assets are digital and optimised for UE, the only logical next step is to consider this a massive creative and business opportunity to produce multi-platform or transmedia content. In simple terms, multi-platform content is the concept of delivering *a story* via different platforms, such as a film, a game, an interactive website, social media, all while conveying the core story. Transmedia also delivers to multiple platforms, but a clear distinction is provided by Moloney who uses crossmedia in place of multi-platform and states that, "Crossmedia = One story, many channels" (2014) in contrast to "Transmedia = One storyworld, many stories, many forms, many channels" (Moloney, 2014). "Transmedia storytelling represents a process where integral elements of a fiction get dispersed systematically across multiple delivery channels for the purpose of creating a unified and coordinated entertainment experience" (Jenkins, 2007). Media creators and producers have harnessed the power of transmedia for years, especially for maximum exploitation of rights of a larger franchise. A recent example of transmedia stemmed from the film *Avatar: The Way of Water* (Cameron, 2022) since it also launched an initiative called "Keep Our Oceans Amazing" which, in essence, has qualities of a metaverse of user-created sea creatures, with the aim of "supporting The Nature Conservancy to protect 10 of our oceans' amazing animals and their habitats" (*Avatar*, 2022). Disney US also donated five dollars per creature made, up to one million dollars. Once a user builds their creature, a link will allow them to persistently follow their water habitant throughout a very lush seascape. This example showcases how a media creators can engage viewers for a large feature film intellectual property (IP) and "expand the potential market for a property by creating different points of entry for different audience segments" (Jenkins, 2007). Whether you are James Cameron, with every advanced technical tool and immense monetary resources, or a student filmmaking team, telling an intimate story dedicated to suicide-bereaved families using free assets from UE (Ibitayo et al., 2021), creators have the choice to embed other avenues *to and from a story*. Best practice for producers is to plan the narrative(s) of a project from the beginning, however, this new pipeline affords flexible opportunities for iteration at later stages. Technically, creators do not have to decide on the transmedia portions immediately or at the time of creating the initial work, since the files remain Unreal Engine files and, even with software updates, can be updated, and perhaps when an idea or funding for an extension for the IP enters into the picture, creators do not have to travel to a far-off setting with a crew who is now on different projects and actors who are booked. A creator can simply open an older UE project file, copy it to a different 'level', only now to produce a game, to give their story a new life in a different form. For media creatives, this is a blessing and a curse since finessing could potentially be infinite. Media creatives must remember that the medium itself will call for how it can be used. As Marshall McLuhan famously stated, " 'The medium is the message', which means in terms of the electronic age, that a totally new environment has been created" (as cited in McLuhan & Gordon, 2003, p. 13). As the meta-verse evolves to become the new environment for teaching subjects like virtual production and architecture, it is critical to consider the inherent opportunities afforded by this massive transformation.

10.4 FILM PRODUCTION EDUCATION IN THE METAVERSE

The concept of a film school in the metaverse may seem controversial to some; however, it has already occurred in various forms for some time. The doctoral dissertation "Developing a Learning Model to Teach Film Production Online" detailed a study with a goal to democratise film production education, analysing how all subjects of film production could be taught online and virtually, and used the example of an originating metaverse, namely, *Second Life*, (Ziemsen, 2017). It was concluded that nearly all subjects spanning development, pre-production, production, post-production and distribution and exhibition could all be taught virtually and online – in the metaverse. This proposed concept was received with some scepticism by film professionals however full comprehension and understanding by those who already worked in real-time animation. Since this time, several virtual, online, and metaverse film schools have come to market, including Virtual Film School, which expected to raise over one million dollars in crowdfunding (Smith, 2022) in 2022 and aims to train, "with live instructors, real mentorship, and using Virtual Reality as the next-level classroom" (*Virtual Film School*, n.d.). Another example recently launched is The Metaverse Film School LLC, which gives students access to "fully immersive, virtual soundstages and multi-camera production studios" (The Metaverse Film School LLC, n.d.) and gives access to 'virtualised' facilities including a motion capture stage, production offices, green screen, and even has options to go back in time to "Thomas Edison's Black Maria – the first movie studio ever built" (The Metaverse Film School LLC, n.d.). The concept of simulated and virtual film education dates back to an application called Moviestorm, which has "been democratising 3d storytelling through the use of accessible and affordable tools since 2005" ("Moviestorm," n.d.) and has evolved product offerings, such as "FirstStage" in the current virtual production ecosystem. Creating virtual applications for filmmaking has been heralded as revolutionary, albeit by few early adopters, who have built prototypes, such as the game *Machinima Film Studio v1.0*, conceived of by faculty and built by a collaborative student team of game and film college students over the course of a semester (Ziemsen, 2014). *Machinima Film Studio v1.0* is a multiplayer game that allows users to 'play' the roles of director, cinematographer, actor, cinematographer, shoot film on a virtual film set, output a film, and watch it together in a final theatre for a red-carpet premiere (Ziemsen, 2014). Another hybrid game/app application that is targeted specifically at cinematographers is called *Cine Tracer*, which is a "real time cinematography simulator made with Unreal Engine" (*Cine Tracer*, n.d.) and enables users to use realistic lighting and camera equipment to practice the art and technique of cinematography virtually. Learning the intricacies of operating a virtual production LED volume/stage will soon be fully democratised with the newly launched multiplayer metaverse prototype built with Unreal Engine, of a virtual "replica of a professional volume . . . to learn about the new pipeline" (Ziemsen & Fenuta, 2022). This application, and eventual interoperable metaverse destination, enables virtual production students to learn about technical components of a LED Volume, but also, screenwriters can have the chance to immerse themselves in nearly any environment from UE Marketplace, to ideate in a story world (Ziemsen & Fenuta, 2022). It is evident that the surge of tools for virtual filmmaking will only increase with advances in the metaverse.

10.5 ARCHITECTURE EDUCATION IN THE METAVERSE

Architects will help shape the metaverse by way of innovations in visualisation software such as Twinmotion, which enables real-time rendering and immersive human-scale models (An Overview of Twinmotion Features and Capabilities, n.d.). Specifically, it interfaces with Unreal Engine and allows for multi-user, remote, collaborative, and persistent experiences of both exterior and interior space with the ability to make changes in real time. The current architectural industry-standard 3D visualisation tools are limited to static 3D imagery and architectural animation which do not allow the user to physically enter spaces until they are built. It is not economically feasible for most students and clients to create full-scale architectural mock-ups prior to construction. Mock-ups are costly and can delay a project once in construction. Simulating architecture in an immersive and persistent manner in addition to allowing many users into a space simultaneously will serve to inform a design and allow for changes that were not possible before. The metaverse offers a parallel virtual experience where the client, architect, professor, and student can collaborate within the digital design to make changes in real time while immersed in the space (Kolata, 2022). Digital twins can be created with the use of new software workflows by bringing building information modelling (BIM) files into Unreal Engine, (Dejtiar, 2022). Creating a digital twin of any given building will provide students a nuanced learning experience where their spatial and understanding of materials and methods construction will all be tested in an authentic and immersive way. As Ball (2022) states, "The use of "game engines" to power such simulations does make it easier to produce a Metaverse which spans both the physical and virtual planes of existence" (p. 118).

The metaverse offers interdisciplinary education of historically separate fields of film and architecture to collaborate in a meaningful, immersive, and virtual setting (Ziemsen & Fenuta, 2022). Architects designing for the metaverse will no longer engage with traditional design teams such as mechanical, structural, electrical engineers, and contractors; rather, they will be engaging with new disciplines such as computer programmers, UX designers, and even product designers (such as VR headsets) (Schumacher, 2022). However, the collaboration with other disciplines is endless as architects can engage with any discipline who wishes to collaborate in the metaverse (Bayrak, 2022). Architects will become benefactors in the emerging digital economy of the metaverse. Both architects and students will have an international stage where conceptual designs can be monetized in the form of NFTs (non-fungible tokens). As stated by Hackl et al. (2022) "An NFT is a digital asset that is unique and singular, backed by blockchain technology to ensure authenticity and ownership. An NFT can be bought, sold, traded or collected" (p. 7). For example, economist and financial analyst Adonis Zachariades co-founded Renovi, a company that creates architectural NFTs and aims to become the marketplace where architects sell their designs across all metaverse platforms (Sheber & Speros, 2022). There will be a low barrier of entry to sell on these platforms, where peer review from critically acclaimed architects will need to exist in order to validate the worth of architectural concepts and the price of the NFT. As Chloe Sun states, "Every designer has amazing designs sleeping in their hard drives. . . . We can revitalise them in the metaverse, where we have more opportunities to share our creativity with the world" (HOK, 2022).

10.6 CONCLUSION

The innovation and adoption of film production, architecture, and the education of these fields in the metaverse has evolved exponentially in recent years, and it is expected to continue to grow in all dimensions – technologically, theoretically, creatively, logistically, and socially. The timeless words of Marshall McLuhan ring true about the current moment, the metaverse goldrush. "Each new technology creates an environment that is itself regarded as corrupt and degrading. Yet the new one turns its predecessor into an art form" (McLuhan & Gordon, 2003, p. 13). "If it works, it's obsolete" (McLuhan & Gordon, 2003, p. 24). Gauging by the many examples of virtual film education and immersive innovations in how to teach architecture, it is evident that we have entered a new dimension of the education in traditionally applied and practical fields. We are nearing a maturity in the metaverse space that is dramatically more accepting of this innovation than even five years ago. Many metaverse participants are still mesmerized by the advent of using headsets to enter virtual, immersive, persistent platforms to facilitate education, in a similar manner to how people physically ran out of movie theatres while watching *L'Arrivée d'un Train en Gare* (depicting a train arriving in a station), which tricked novice moviegoers into thinking that the train would come charging through the screen (Cook, 1996). The ecosystem of education is starting to become more open to flexible and virtual learning platforms and tools, including writing it into their academic plans (*Academic Plan: 2023–2026 – Looking Forward*, 2022). Film production, which has evolved in great extent to virtual production, is primed and ready for harnessing the power of Unreal Engine and other tools to revolutionize a medium that is ripe for a transmedia approach. Like film, the education of architecture has the potential to enable professionals and educators alike to reduce barriers and costs like never before using visualization software like Twinmotion. Filmmakers and architects can collaborate with each other and almost any other medium, in ways that were never previously possible. The metaverse is a new junction point, which will enable infinite, exponential, vast, and equally intimate and personal projects, narratives, experiences, films, and designs. In practical terms, it can be a great premise to bring faculty members together with fields and colleagues that, historically, they may not have interacted with, such as game programming, film production, virtual production, architecture, and interior design (Ziemsen & Fenuta, 2022).

REFERENCES

Academic Plan: 2023–2026 – Looking Forward. (2022). Humber College. https://humber.ca/academic-division/sites/default/files/documents/Academic_Plan.pdf

Avatar. (2022). Retrieved December 16, 2022, from www.avatar.com/KeepOurOceansAmazing

Bada, D., & Olusegun, S. (2015). Constructivism learning theory: A paradigm for teaching and learning. *Journal of Research & Method in Education, 5*(6), 66–70.

Ball, M. (2022). *The Metaverse: And how it will revolutionize everything.* Liveright Publishing Corporation, a division of W.W. Norton & Company.

Bates, T., & Poole, G. (2003). *Effective teaching with technology in higher education.* Jossey-Bass/John Wiley.

Bayrak, S. (2022, July 11). Is metaverse really the end of barriers for architects? *ArchDaily.* Retrieved December 9, 2022, from www.archdaily.com/984891/is-metaverse-really-the-end-of-barriers-for-architects

Cameron, J. (Director). (2022, December 16). *Avatar: The way of water* [Action, Adventure, Fantasy]. 20th Century Studios, TSG Entertainment, Lightstorm Entertainment.

Cicek, I., Bernik, A., & Tomicic, I. (2021). Student thoughts on virtual reality in higher education – a survey questionnaire. *Information, 12*(4), 151. https://doi.org/10.3390/info12040151

Cine Tracer. (n.d.). Cine Tracer. Retrieved December 22, 2022, from www.cinetracer.com

Cook, D. A. (1996). *A history of narrative film.* W.W. Norton.

Dejtiar, F. (2022, November 26). The architecture of the metaverse (so far). *ArchDaily.* Retrieved December 9, 2022, from www.archdaily.com/988639/the-architecture-of-the-metaverse-so-far

Goldsmith, J. (2022, February 1). Spire animation studios closes $20m funding round with strategic investment by epic games. *Deadline.* https://deadline.com/2022/02/spire-animation-epic-games-danny-mcbride-trouble-caa-1234923799/

Hackl, C., Lueth, D., & Bartolo, T. D. (2022). *Navigating the metaverse: A guide to limitless possibilities in a web 3.0 world.* Wiley.

Harwood, Tracy, & Grussi, B. (2021). *Pioneers in machinima: The grassroots of virtual production [hardback].* Vernon Press.

HOK. (2022, February 28). How architects and designers can help define the metaverse. *HOK*. Retrieved December 9, 2022, from www.hok.com/ideas/research/how-architects-and-designers-can-help-define-the-metaverse/#:~:text=HOK%20has%20long%20used%20virtual,and%20play%20out%20various%20scenarios.

Ibitayo, O., Del Pilar, J., & Kharshina, N. (2021). *The lives left behind*. https://game-fmst-collaboration.webflow.io/film/the-lives-left-behind

Infogence Global Research. (2021). *Global virtual production market (2021–2026) by component, type, end-user, geography, competitive analysis and the impact of covid-19 with ansoff analysis*. www.researchandmarkets.com/reports/5544140/global-virtual-production-market-2021-2026-by

Jenkins, H. (2007, March 21). Transmedia storytelling 101 – pop junctions. *Henry Jenkins*. http://henryjenkins.org/blog/2007/03/transmedia_storytelling_101.html

Johnson, P., & Pettit, D. (2012). *Machinima: The art and practice of virtual film-making*. McFarland & Company.

Kolata, S. (2022, April 5). Metaverse vs. sustainability: How can the metaverse help us deliver better designs? *ArchDaily*. Retrieved December 9, 2022, from www.archdaily.com/979750/metaverse-vs-sustainability-how-can-the-metaverse-help-us-deliver-better-designs

Marino, P. (2004). *3D game-based filmmaking: The art of machinima*. Paraglyph Press.

McLuhan, M., & Gordon, W. T. (2003). *Understanding media: The extensions of man – critical edition*. Gingko.

The Metaverse Film School LLC. (n.d.). *Metaverse Film School*. Retrieved December 22, 2022, from www.metaverse-filmschool.com/what-we-do

The Metaverse Standards Forum. (2022). *Metaverse Standards Forum*. Retrieved October 27, 2022, from https://metaverse-standards.org/

METAVERSE X APG. (2023). Retrieved May 3, 2023, from https://metaverse-x-apg.webflow.io/

Milmo, D. (2021). A whole new world: Disney is latest firm to announce metaverse plans. *The Guardian*. www.theguardian.com/film/2021/nov/11/disney-is-latest-firm-to-announce-metaverse-plans

Moloney, K. (2014, April 21). Multimedia, crossmedia, transmedia . . . what's in a name? *Transmedia Journalism*. https://transmediajournalism.org/2014/04/21/multimedia-crossmedia-transmedia-whats-in-a-name/

Moviestorm. (n.d.). *Moviestorm FirstStage*. Retrieved December 22, 2022, from https://firststage.moviestorm.co.uk/cover-page/

An Overview of Twinmotion Features and Capabilities. (n.d.). Twinmotion. Retrieved October 5, 2022, from www.twinmotion.com/en-US/features

Piaget, J. (1952). *The origins of intelligence of children*. International Universities Press.

Schumacher, P. (2022). The metaverse as opportunity for architecture and society: Design drivers, core competencies. https://doi.org/10.1007/s44223-022-00010-z

Sheber, A., & Speros, W. (2002 September 14). The hotel industry enters the Metaverse. *Hospitality Design: HD* [Online]. https://hospitalitydesign.com/news/development-destinations/hotel-industry-nfts-metaverse/

Smith, B. (2022). World's first metaverse film school announces crowdfunding launch at YA partners. *EIN Presswire*. www.einpresswire.com/article/560977455/world-s-first-metaverse-film-school-announces-crowdfunding-launch-at-ya-partners

Storyliving™ by Disney. (2022). *Storyliving™ by Disney*. Retrieved October 4, 2022, from www.storylivingbydisney.com/

TEDx Talks (Director). (2014). Virtual reality – How the metaverse will change filmmaking | George Bloom. *TEDxHollywood*. www.youtube.com/watch?v=ZjwjomAPMlw

Terry, Q. H., Skee, K. S. D. J., & Hilton, P. (2022). *The metaverse handbook: Innovating for the internet's next tectonic shift*. Wiley.

Truong, D. (2022, October 10). Overwhelming demand for online classes is reshaping California's community colleges. *Los Angeles Times*. www.latimes.com/california/story/2022-10-10/huge-online-demand-reshapes-california-community-colleges

Unreal Engine Marketplace | Store of UE Assets for Games and 3D Rendering. (n.d.). Retrieved January 4, 2023, from www.unrealengine.com/marketplace/en-US/store

Virtual Film School. (n.d.). Retrieved December 22, 2022, from https://virtual-filmschool.org/

Zhao, Y., & Frank, K. (2003). Factors affecting technology uses in schools: An ecological perspective. *American Educational Research Journal*, 40(4), 807–840.

Ziemsen, E. (2017). *Developing a learning model for teaching film production online*. University of British Columbia. https://open.library.ubc.ca/collections/ubctheses/24/items/1.0361756

Ziemsen, E. (Director). (2014). *Machinima Humber College 2014*. www.youtube.com/watch?v=thgLZtvlnk0

Ziemsen, E., & Fenuta, E. (2022). *Harnessing Higher Education in the Metaverse A Collaboration between Architecture and Virtual Production (film)*. Harnessing Higher Education in the Metaverse – A Collaboration between Architecture and Virtual Production (Film). https://metaverseedu.webflow.io/

CHAPTER 11

Gone with the Wind

From Virtual Reality to the Metaverse at Film Festivals

Dr. Montserrat Jurado-Martín

Universidad Miguel Hernández, Elche, Spain

CONTENTS

11.1 INTRODUCTION

Film festivals are a reflection of trends, and they directly or indirectly opt for innovation in all areas, techniques and formats, including virtual reality, augmented reality, immersive reality, 360°, and, ultimately and most novel, the metaverse. Moreover, regarding film festivals as cultural industries, the academic literature is almost anecdotal compared to that related

DOI: 10.1201/9781003379119-11

to, for example, video games or experimental art, even though film is probably one of the most creative media.

During the 2022 edition and in the context of the Cannes Film Festival, the first NFT Film Festival took place, a full metaverse experience located in a virtual place called Decentraland. James Ellis (2022) stated that this event showed the potential of the metaverse while offering workshops, conferences, and networking, thus providing access to Cannes through the NFT Film Festival and offering a meeting space for creators and investors in the sector.

In this context we contemplate the intersection among film festivals, virtual reality, augmented reality, immersive reality, 360°, and the metaverse. The study on virtual reality film festivals (Jurado-Martín, 2021) concluded that, despite being of interest to the film and audiovisual market, there were barely a dozen specialised or exclusive events. This is a niche that has a long way to go. That work left the door open to the study of specific sections or activities without the need for them to be specialised. This chapter analyses pioneering experiences in the use of metaverse techniques at film festivals, tries to differentiate these techniques from all the previous ones (virtual reality, augmented reality, immersive reality, 360°) and to discover whether this could really be a future trend or just a passing fad. The metaverse has arrived on the scene of new technologies like a hurricane, overshadowing others that are not yet completely consolidated. These circumstances deserve further analysis and reflection.

11.2 EXPLORING NEW NARRATIVES AND TECHNIQUES IN THE FILM FESTIVAL ENVIRONMENT

Cinema is probably one of the most creative and innovative media. "Cinema is more than a means of communication", explains Jerónimo Repoll (Becerra, 2019) applying McLuhan's axiom (1964) "cinema is the message". While print or digital media, radio or television are linked to journalistic functions and, consequently, to a greater commitment to informative routines, the film medium moves more freely in the artistic field.

From this starting point, we could assume that the inclusion of innovative formats is more accessible to cinema. However, very little has been written about cinema in relation to the use of virtual reality. In the audiovisual environment, these formats have been more developed for experimental content rather than fiction (Martínez-Cano, 2018, pp. 167–169) and for video games.

Virtual reality began to be used in the field of video games as a tool aimed at immersing the player and enhancing their active role. This is how "virtual reality questions cinema and challenges the audiovisual language and its convention when constructing narrative strategies" (Martínez-Cano, 2015, p. 163). This author states that an immersive experience is generated between the spectator and the event. In the same vein, Cortés-Selva (2015, pp. 265–266) describes that immersivity reaches its highest expression in 3D cinema. She thus follows the approach of other authors such as Murray (1999) who contemplated immersivity in virtual environments and the possibilities of user involvement.

"The application of virtual reality techniques in cinema must be reflected in film festivals and in the way they promote this format" (Jurado-Martín, 2021). This reflection can be seen in three active participants: producers/directors, festival organisers, and audience. In the case of metaverse experiences at film festivals, a new factor has emerged: technology companies. They see festivals as a space for business and promotion.

Film festivals are non-commercial events from which creators spread and promote their most innovative works (Jurado-Martín, 2006). They are "key spaces for production, dissemination, exhibition and exchange" (Peirano, 2018, p. 66). Furthermore, they were the first to accept digital format as opposed to 35 mm film, democratising access to audiovisual creation for many young talents. In this context, film contests are a reflection of trends, and they directly or indirectly opt for innovation in all areas: ideas and stories in the scripts, techniques used, and formats.

Without claiming to be exhaustive about the studies that relate cinema and virtual reality, we would like to highlight the works by Gifreu-Castells (2017, p. 2) on the consumption of this type of film and the transformation into the individual experience; Sementille et al. (2014, p. 93) and the applications of augmented reality in television; and especially all those authors focusing on the transformation this has represented in the field of virtual reality, such as Rose (2011), Sherman and Craig (2002), Harbord (2002), Holmberg (2003), Gantier and Bolka (2011), Boas (2013), Cuadrado Alvarado (2014) or Jurado-Martín (2020, 2021); and in the field of journalism, such as De la Peña et al. (2010), Domínguez (2012), Paíno et al. (2017), among others. Contextual reference to the study of film festivals by De Valck (2007) and De Valck et al. (2016) is most definitely needed (Iordanova & Leshu, 2012).

The festivals with the greatest prospects are those which provide a space for dissemination specializing in a specific area, format or content; those that, regardless of their offerings, have public or private support that goes beyond what is strictly cultural; and finally, those with both characteristics. The symbiosis between metaverse and film festivals can be found in this interaction. Film festival organisers may be interested in the offerings provided by a specialised contest, where new technologies and cinema are combined. Indeed, a growing number of researchers are emphasising this value, focusing attention not only on large international festivals but also on the impact that smaller festivals can have on the film sector economy and on their own environment as a cultural or social force.

11.3 PRE-METAVERSE EXPERIENCES AT FILM FESTIVALS

The film festivals with a metaverse experience being considered are those employing techniques and practices of virtual reality, 360° cinema, and immersive cinema. According to the results of Jurado's study (2021), these specialized, exclusive events have in common that they are essentially new, as nearly all of them have less than six editions, and that they are mainly held in the last quarter of the year (71%, out of which 42% in September). It is remarkable that no country has two or more of such events. The American continent is represented by three (43%) of these festivals and Europe by four (57%).

In general, these types of contests are international in nature, open to participants from all over the world, with a general content (86%) and with a preference for short films (71%). They all have their competition rules published, although incompletely in some cases (29%), and include the list of awards and a description of the same in said rules. All of them – except for one – have several award categories, and a very few of them award cash awards – only two festivals, 29%. They all describe the registration procedure in their rules, and there is a balance between those festivals which are free (Mexico, Germany, and Italy) and those which include a fee (USA, Canada, and France).

In the same study, 12 out of the 19 festivals resulting from an initial filtering of the FilmFreeWay (FFW) distribution platform database were discarded on the grounds that they were not considered film festivals (one music festival, one project competition, one film competition, and four film exhibitions, as well as five other festivals that were no longer active). Therefore, taking as a sample the more than 7,000 film festivals on the FFW distribution platform, only seven film festivals were found to specialise exclusively in virtual reality, immersive cinema, and 360° cinema at the international level.

Given the small number of this type of film festivals, it can be stated that there is no interest in organising virtual reality and related festivals. In fact, it was observed that, in order to avoid expenses, the tendency is for these events to be rather fee-paying exhibitions, where prizes are reduced, and no parallel activities are organised. This scarce number of specialized festivals could be related to the fact that their potential audience is a minority and that the exhibition resources are greater than those for traditional festivals.

Since organisers do mark the virtual reality category when registering the event, it could be considered that this is a format of interest but not enough as to organise a festival exclusively on the subject.

11.4 PIONEERING FILM FESTIVALS IN THE METAVERSE ENVIRONMENT

It is too early to speak about film festivals specialising in metaverse since they are hardly representative in terms of number. The events with a tendency to apply these techniques and narratives organise conferences, round tables, markets, and showrooms in which they show the most innovative trends and provide them to industry professionals. The way to get to know these activities is the dissemination by the events themselves through the media or in person.

11.4.1 The Debate about Metaverse at Film Festivals

Experiences on the metaverse are still practically anecdotal, and, in some cases, they do not go beyond a declaration of intent or a headline. In the case of Seriesland, an International Webseries Festival, Elorriaga (2022) underlined the strong influence received from metaverse and virtual reality in the audiovisual productions in competition. However, it was only a catchy headline.

The Sitges Horror and Fantastic Film Festival received media coverage with headlines announcing that the event was opening its doors to the metaverse. However, in the end, all the focus was on the out-of-competition screening of the film *Tron* by Steven Lisberger, coinciding with the film's 40th anniversary (EFE, cited in the newspaper *La Vanguardia*, 2022). Elperiodico.com (2022) headlined the same news item using the term 'virtual reality' and asserted the following: "When *Tron* was released, nobody talked about metaverse, where physical reality is mixed with virtual reality. . . . Nobody could imagine its influence and how the concept of virtual reality would develop in these four decades". This statement makes clear the indiscriminate use of the terms 'metaverse' and 'virtual reality' to refer to the same concept.

The Santander Film Festival[1] organised a round table attended by Spain Audiovisual Hub advisor Francisco Asenci and other collaborators. They concluded as follows:

> The metaverse continues to mature in order to find its place in the audiovisual sector. New forms of content need to be regulated so that they can be efficiently and safely exploited. The film industry must do its part to adapt to new formats and to be able to embrace the possibilities offered by new technologies.

There is no doubt that preliminary essays are needed to spotlight both the existence of these experiences and the interest in their study.

The San Sebastian Film Festival, the most international Spanish film festival, organised a meeting to debate about the metaverse.[2] Under the confusing heading *El metaverso que es, no es, pero al final acabará siendo* [*The Metaverse that is, is not, but in the end will end up being*], Elorza (2022) precisely conveys the indeterminacy of the metaverse concept, which is in the process of self-definition. The author recalls Kim Magnusson's statement that "the metaverse is the next step of something we already had", as opposed to Beatriz Pérez de Vargas' statement affirming that "the metaverse does not exist, it is about countries" in the context of the debate on the need to legislate a technology that goes beyond borders. The text opens and closes a debate questioning the metaverse approaches as the revolution of an unprecedented phenomenon:

> According to Beatriz Pérez de Vargas, the metaverse is an umbrella term encompassing many realities: "Human beings have been telling stories for 20,000 years with the aim of generating feelings; with the arrival of the metaverse, we will no longer be mere storytellers but rather let the user live these stories". Instead of going online, we will live online. So, what will future content be like? Interactive, immersive and personalised. Everyone will be able to create content. This poses a challenge that could become a threat: uniformity in the audiovisual world. Nothing new in the old universe.
>
> (Elorza, 2022).

11.4.2 Film Festivals Specialising in the Metaverse

This context does not encourage a detailed scientific analysis since there is little more than the study of isolated cases with a great heterogeneity of proposals. Nevertheless, this chapter develops a preliminary study of film festivals specialising in the metaverse. The results are shown later.

Taking the FFW database as a reference, by entering the terms 'virtual reality', 'augmented reality', and '360° cinema' and then adding the filter 'film festival', we obtain a total result of 984 events.[3] We assume that many of these events are repeated, others are not film festivals but another type of event, some of them are no longer held, and there is also a percentage of events which are questionable in terms of celebration due to the absence of contest rules or to the presence of incomplete rules. Given this data, by entering the term 'metaverse' and the filter 'film festival', we obtain a result of five festivals: Belfast XR Festival (UK, first edition in 2022), New Media Film Festival (USA, first edition in 2009), Splice Film Fest (USA, first edition in 2017), Orlando Urban Film Festival (USA, first edition in 2014) and FilmGate Interactive Festival (USA, first edition in 2014).

Belfast XR Festival[4] combines in its description the use of immersiveness (I) and virtual reality (VR). They state that their aim is the dissemination of VR and AR (augmented reality) content. This festival is international in nature, but they value the presence of local participation. There is a registration fee of £10. There are no cash awards, and there is only one in-kind award consisting of the coverage of expenses to offer a workshop on the topic. Belfast XR Festival accepts 360°, VR, and AR video projects on any subject.

New Media Film Festival,[5] based in Los Angeles (USA), is described as an innovative contest that provides a catalytic environment for the evolution of audiovisual technology. One advantage of this festival is the possibility for participating productions to be considered for Netflix content. It has a metaverse category, in addition to 30 other categories: 360°, AR, AI, 5D, Animation, Artwork, Digital Comics, Documentary, Drone, Faith & Family, Feature, Gaming, Mixed Reality, Mobile/Tablet, Music Only, Music Video, NFT, New Media, Pilots, Pitching, Podcasts, Scripts, Shorts, SR- Socially Responsible, STEAM, Student, Trailers, TV, Virtual Reality, and Web Series. It offers more than $45,000 in in-kind awards.

Splice Film Festival[6] is based in New York and specialises in film and video art. It showcases non-commercial short films and music videos ranging in length up to 10 minutes. They accept many different works, such as abstract/experimental, video art, animation, augmented/mixed reality, documentary, traditional narrative, comedy, social justice, LGBTIQQ, mixed media, music video, performance art, among others. The works can be professional or amateur. There are 20 categories, two of which are specific to the concerned subject: Best AR/Mixed Reality and Best Student AR/Mixed Reality. However, there is no specific category for the metaverse concept.

Orlando Urban Film Festival (OUFF)[7] is an event specialising in new narratives and techniques of African American cinema. There are several registration categories, where the metaverse is differentiated from other categories such as VR or AR. Those interested can submit their work for the following sections: Feature Films, Short Films, Clips, Avatar Images and Clips, Metaverse Links, Metaverse Videos, AR/VR Projects, Gaming Videos, Tic-Tok Videos and Screenplays. There are 13 award categories. According to the description on their website: "OUFF's key objective is to help Content Creators display their works on a big screen or perform live on large venue stages and in our Metaverse Spaces". They also point out the following: "We provide Content Creators and their works with much-needed media exposure to help advance their recognition in the entertainment Marketplace".

OUFF organisers strongly emphasise that "OUFF is where Urban Movie, Television, Music, and Internet Content Producers come together to join forces to discuss the collaboration, dynamics, and interconnection of images, audio, and music". They also state that "Content Creators are Visionaries in movie, music, art, and technology". However, all this enthusiasm clashes with the simplicity of their website despite the many editions that have been held. The value they give to the internet for the promotion of creators and their productions is not matched by the content included on the website's three tabs: 1. home: with non-dynamic images and a poor description, 2. link to last edition's schedule (a four-page pdf), and 3. link to ticket sales.

FilmGate Interactive Festival[8] is described as "an Interactive Media Festival exploring how emerging technology, empowers the content of the future". The event is held in Miami, and since their launch in 2014, they have "hosted fire chats, labs, showcased the works of more than 300 creatives, and co-produced over 20 creative projects" with originality and innovation as common features. "As we approach a united virtual space at increasing rate, aka 'the metaverse', we explore the multiverses and 3D spaces that are paving the roads to the future", they stated.

The awards are as follows: Best Interactive Screening Narrative – Interactive, immersive or cross-platform project, which employee's new tech in the most innovative and organic way to tell a great story; Best Interactive Documentary – Interactive, immersive or cross-platform project. This year we are especially interested in projects on the following themes: science and art, artificial intelligence, environmental conservation, gun control; Best Interactive Exhibit – Interactive, immersive or cross platform project, deliverable and participant-ready; Best Interactive Music event – for the best immersive or cross platform event, where music is the driving forcé; and Audience award. No monetary amount is specified.

This contest explicitly specifies what they consider to be an interactive project and gives some examples: "Projects must be interactive or immersive or be told across multiple platforms. Nonexclusive examples of such stories are: web projects, mobile apps, games, multi-media installations and multi-platform Works".

11.4.3 Other Remarkable Experiences on Metaverse at Film Festivals

Although the FFW platform gathers many festivals, the 'metaverse' concept is so recent that it is possible for many organisers not to contemplate the same on their websites or that platforms are out of date, as is the case of the Sundance Film Festival.[9] For this reason, and as a case study, two European experiences are analysed: Venice International Film Festival and Cannes Film Festival, as well as a brief mention of the above-mentioned Sundance Festival.

The French Rivera Film Festival[10] collaborates with Sundance's actions and praises the metaverse projects they are developing, summarising them as follows:

> The 2022 Sundance Film Festival is kicking off nine days of film premieres, conversations, and immersive experiences. Event organizers have partnered with digital production studio Active Theory to create a spaceship and bio digital showcase via VR headset. That's not the only VR option for attendees. Use customizable avatars to explore these three virtual worlds during the film festival: New Frontier Gallery – A digital event space where attendees can browse the entire catalogue of XR content available as part of the New Frontiere showcase and chat with attendees and festival employees via proximity chat. Cinema House – An immersive

stage environment that will serve as the primary venue for various screenings throughout the festival. Film Party – A social space where attendees can chat with Sundance creators about their respective projects using proximity audio and video chat.

The Biennale Venice International Film Festival[11] has two sections on virtual reality: a specific film programming and a space for social events and talks. Venice VR Expanded offers talks, concerts, guided tours, and so forth, under the concept of virtual reality. These sections appeared in 2021. In the case of the virtual reality category, the first reference was in 2017. Three subcategories are awarded, but the monetary amount is not specified. Up to 30 productions and ten out-of-competition productions can be exhibited. However, as explained by its organisers in the article "Venice Film Festival Announces Metaverse Entry With 'Venice Immersive'" (Egede, 2022), the absence of a market focused on these practices hinders its development and relegates it to a secondary activity where curiosity and passion for the seventh art prevail.

> Over the past two years, the Venice Film Festival has evolved to include a virtual program to meet the demands of diverse audiences. It has adopted a more inclusive: "Venice Immersive." The focus is not only on VR technology but also on including other diverse things to produce an immersive experience for users. . . . The goal is to diversify the metaverse landscape to achieve more hybrid results. . . . However, the lack of a real immersive art market will hinder the team's drive. Venice creators do it out of curiosity and passion for the art.
>
> (Egede, 2022)

Finally, we refer to the metaverse experience of the Cannes Film Festival.[12] Their participation rules do not specify the possibility of submitting specific material on virtual reality. They only mention one award for the best feature film and another award for the best short film, not indicating the monetary amount. The official website search engine does not return any results for the word 'metaverse'. However, there are 30 results when the word 'virtual reality' is entered: two films, 19 photographs, seven news items, and one press release, but almost none of them are related to virtual reality activity.

Virtual reality can be found at Marché du Film, within the Discover section, in the Cannes XR and NEXT areas. Cannes XR is described as the following:

> It is a program from the Marché du Film dedicated to immersive technologies and cinematographic content. It is the unmissable annual *rendez-vous* for XR community offering a multitude of networking, financing, and distribution opportunities. It is a networking platform with which directors, studio executives, XR artists, independent producers, leading tech companies, location-based and online distributors all gather to discuss the role of XR technologies, inspire the art of storytelling and fuel the future of film.

Cannes Next is described on the website as the following:

> It is a unique gathering enhancing partnerships and fostering business opportunities by connecting world-class creativity with cutting-edge business and technological innovation. They are inspiring conferences, keynotes, and panel discussions; to grow your network with creatives, clients, and tech companies, among our various events. . . . Cannes Next network with executives and entrepreneurs, attend startup pitching sessions, and explore new business opportunities.

However, it does not directly mention metaverse or virtual reality.

In both cases, the Cannes Film Festival, and the Venice Film Festival, metaverse content is not identified as such, but it is still referred to as virtual reality. These are non-competitive sections aimed at technology and business.

11.5 CONCLUSIONS: METAVERSE AND FILM FESTIVALS: A FUTURE TREND OR A PASSING FAD?

The starting point was the assumption that cinema is the medium closest to art, and, for this reason, it has a greater interest in the application of experimental techniques with new technologies. As can be inferred from the scientific literature, the use of virtual reality, augmented reality, 360°, and immersive cinema in the film industry is mainly applied in the fields of art, animation, and documentaries, but it also finds a place in fiction.

Film festivals are a key space for the dissemination of the most innovative and creative proposals. It can therefore be deduced that the application of virtual reality techniques in the film industry should be reflected in film festivals through the evolution in the use of techniques, formats, and narratives. This reflection is twofold: The interest of producers and creators in disseminating their projects and the interest of festival organisers in disseminating such projects.

This chapter has analysed several pioneering experiences in the use of metaverse techniques at film festivals. However, it has not been possible to draw a clear boundary among virtual reality, augmented reality, immersive reality, 360°, and the new approach: The metaverse. The latter is presented as a logical evolution, the result of the sum of all previous approaches, and like a new creature, it is baptised metaverse.

At present, metaverse is mainly found in parallel and non-competitive sections and, more specifically, in those sections related to market and networking. These are spaces that are in the interest of technology companies in the sector, but they do not have a great artistic repercussion worldwide, nor do they represent a competitive category of film festivals.

The Lumière's invention was the sum of other creations that gave rise to something new: the cinematograph. The metaverse is undoubtedly a new reality with its own technique and narrative. It is based on entertainment, curiosity, and love of art, as evidenced by film festivals, which are barometers of cinematographic trends. Although the metaverse owes its origins to all the previous formats, it has a unique component of its own. The same happened to cinema in its early days, and today there is no doubt about the revolution it represented as a means of social communication.

We are reaching a point where we will no longer talk about VR, AR, 360°, or immersive cinema but only about the metaverse. Because all previous concepts will be gone with the wind. It is only a matter of time.

NOTES

1 Full text in https://bit.ly/3YyBgkY
2 Conclusions published on the event's website: https://bit.ly/3YwuxYO
3 Distributed into 832, 2, 425 y 159, respectively
4 Official website: www.belfastxrfestival.com/
5 Official website: www.newmediafilmfestival.com/
6 No official website
7 Official website: www.orlandouff.com/
8 No official website
9 Organising company's website: www.sundance.org/

10 Full content at: https://frenchrivierafilmfestival.com/the-metaverse-and-the-future-of-filmmaking/. They do not undertake any activity in relation to the metaverse.
11 Official website: www.labiennale.org/en/cinema/2023
12 Official website: www.festival-cannes.com/es/

REFERENCES

Becerra, A. (2019). Entrevista a Jerónimo Luis Repoll: Hay que pensar el cine como algo más que un medio de comunicación. *Revista Mexicana de Comunicación, 144.* https://tinyurl.com/y28rjj5x

Boas, Y. (2013). Overview of virtual reality technologies. *Interactive Multimedia Conference.* shorturl.at/DMSV9

Cortés-Selva, L. (2015). Viaje al centro de la inmersión cinematográfica. del cine primitivo al VRCinema. *Opción, 31*(4), 352–371.

Cuadrado Alvarado, A. (2014). Tocar a través del cuadro: una genealogía del interfaz como metáfora de control en el espacio del arte, el cine y los videojuegos. *Icono 14, 12*(2), 141–167. https://doi.org/10.7195/ri14.v12i2.708

De la Peña, N., Weil, P., Llobera, J., & Giannopoulos, E. (2010). Immersive journalism: Immersive virtual reality for the First-Person Experience of News. *Presence: Teleoperators and Virtual Environments, 19*(4), 291–301.

De Valck, M. (2007). *Film festivals: From European geopolitics to global cinephilia.* Amsterdam University Press.

De Valck, M., Krendell, B., & Loist, S. (2016). *Film festivals: History, theory, method, practice.* Routledge.

Domínguez, E. (2012). *Periodismo inmersivo: fundamentos para una forma periodística basada en la interfaz y la acción.* Tesis doctoral Facultat de Ciències de la Comunicación de Blanquerna, Repositorio. shorturl.at/cdgUZ

Egede, I. (2022, de agosto 31). Venice film festival announces metaverse entry with 'venice immersive'. *Crypto.news.* https://bit.ly/3FyGG6O

Ellis, J. (2022, de mayo 19). The first NFT film festival in the Metaverse to launch in decentraland. *The Mediaverse.* https://bit.ly/3BMmEEI

Elorriaga, G. (2022, de octubre 13). El festival Seriesland analiza la irrupción del metaverso. *El Correo.* https://bit.ly/3Ywc7r6

Elorza, I. (2022, de septiembre 23). El metaverso que es, no es, pero al final acabará siendo. *Diario del Festival. Festival de Cine de San Sebastián.* https://bit.ly/3PFKRm8

Elperiodico.com (2022, de octubre 1). *Tron y los fascinantes orígenes del cine de realidad virtual en el cine.* https://bit.ly/3YK5jq2

Gantier, S., & Bolka, L. (2011). L'experiènce immersive du web documentaire: études de cas et pistes de réflexion. *Les Cahiers du Journalisme, 22/23,* 118–123. shorturl.at/eitMQ

Gifreu-Castells, A. (2017, Diciembre). Documental interactivo: dispositivos locativos, impacto social, ciencia y divulgación. *Doc On-line,* 2–4. shorturl.at/aepX1

Harbord, J. (2002). Film festivals: Media events and the spaces of flow. In J. Harbord (Ed.), *Film cultures* (pp. 59–75). Sage.

Holmberg, J. (2003). Ideals of immersion in early cinema. Cinémas: revue d´études cinématographiques. *Cinémas: Journal of Film Studies, 14*(1), 129–147.

Iordanova, D., & Leshu, T. (Eds.) (2012). *Film festival yearbook 4: Film festivals and activism.* St. Andrews Film Studies.

Jurado-Martín, M. (2006). *Los festivales de cine en España. Incidencia en los nuevos realizadores y análisis del tratamiento que reciben en los medios de comunicación.* Tesis doctoral Universidad Complutense de Madrid, Repositorio: http://eprints.ucm.es/7306/

Jurado-Martín, M. (2020). Festivales de cine especializados en realidad virtual en España. *Sphera Publica, 1*(20), 59–77. https://sphera.ucam.edu/index.php/sphera-01/article/view/381

Jurado-Martín, M. (2021). Festivales de cine de realidad virtual: aproximación al catálogo de eventos y tendencias de la industria. In M. Viñarás Abad, A. Gregorio Cano, & R. Casañ Pitarch (coords.), *Lo audiovisual bajo el foco del siglo XXI.* Editorial Tirant Lo Blanch.

La Vanguardia (2022, de junio 21). *Festival de Sitges apuesta por mundos virtuales en edición de homenaje a Tron.* https://bit.ly/3jglDi1

Martínez-Cano, F. J. (2015). Nuevos paradigmas del cine y realidad virtual: aplicación de nuevas tecnologías de realidad virtual en videojuegos y cine. In D. Alonso, I. Martínez de Salazar, & J. Cuesta (Coord.), *03 Videojuegos e industria creativa* (pp. 269–282). Escuela de Diseño, Innovación y Tecnología.

Martínez-Cano, F. J. (2018). Impresiones sobre Carne y Arena: práctica cinematográfica y realidad virtual. *Miguel Hernández Communication Journal, 9*(1), 161–190. http://dx.doi.org/10.21134/mhcj.v0i9.222

McLuhan, M. (1964). *Understanding media: The extensions of man.* McGraw-Hill.

Murray, J. H. (1999). *Hamlet en la holocubierta. El futuro de la narrativa en el ciberespacio.* Paidós Multimedia.

Paíno, A., Jiménez, L., & Rodríguez, M. I. (2017). El periodismo inmersivo y transmedia: de leer la historia a vivirla en primera persona. In J. Herrero & C. Mateos (Coord.), *Del verbo al bit* (segunda edición, pp. 1177–1191). La Laguna University. https://doi.org/10.4185/cac116edicion2

Peirano, M. P. (2018). Festivales de cine y procesos de internacionalización del cine chileno reciente. *En Cuadernos.info, 43,* 57–69. https://tinyurl.com/yxkfrf4v

Rose, F. (2011). *The art of immersion. How the digital generation is remaking Hollywood, Madison Avenue and the way we tell stories.* W.W. Norton & Company.

Sementille, A. C., Américo, M., Marar, J. F., & Cunha, A. K. (2014). Arsstudio. Estúdio Virtual para Produção de Conteúdos Audiovisuais em Realidade Aumentada para TV Digital. *Revista Tram(p)as de la comunicación y de la cultura, 77,* 89–98. shorturl.at/epS09

Sherman, W., & Craig, A. (2002). *Understanding virtual reality: Interface, application, and design.* Morgan Kaufmann.

Index

Note: Page numbers in italics indicate a figure or photo and page numbers in bold indicate a table on the corresponding page.

Printed in the United States
by Baker & Taylor Publisher Services